CO
THE EASY WAY
Second Edition

By: Craig E. "Buck," Amateur Radio Call Sign K4IA

ABOUT THE AUTHOR: Known on the air as "Buck," he received his first Amateur Radio license in the mid-sixties as a young teenager. Today, he holds an Amateur Extra Class Radio License.

Buck is an active instructor and a volunteer Amateur Radio examiner. The Rappahannock Valley Amateur Radio Club named him the Elmer (Trainer) of the Year three times.

Email: k4ia@EasyWayHamBooks.com

Published by EasyWayHamBooks
130 Caroline St. Fredericksburg, Virginia 22401

Easy Way books by Craig Buck are available from Ham Radio Outlet stores, Ingram, and Amazon:
Pass Your Amateur Radio Technician Class Test
Pass Your Amateur Radio General Class Test
Pass Your Amateur Radio Extra Class Test
Pass Your General Radiotelephone Operators Test (GROL)
How to Get on HF
How to Chase, Work & Confirm DX

Copyright ©2022, Craig E. Buck All Rights Reserved. No part of this material may be reproduced, transmitted, or stored in any manner, in any form, or by any means without the author's express written permission. 1.4

ISBN 979-8-9856739-0-6

PREPPER COMMUNICATIONS THE EASY WAY
Second Edition

TABLE OF CONTENTS

- TABLE OF CONTENTS 1
- INTRODUCTION 4
- COMMUNICATION FOR PREPPERS 5
- THE IMPORTANCE OF PREPARATION 6
- FREQUENCY 7
- MODES 9
  - FM MODE 9
  - AM MODE 9
  - SINGLE SIDEBAND MODE (SSB) 9
  - PACKET MODE 10
  - CW MODE (MORSE CODE) 10
- POWER 11
  - VOLTS, AMPERES, AND RESISTANCE 11
  - POWER IN WATTS 12
  - GENERATORS 13
  - BATTERIES 18
  - POWER SUPPLIES 24
  - RECOMMENDATIONS 27
- PROPAGATION 28
  - WHERE DO YOU NEED TO COMMUNICATE? 31
- RECEIVERS AND SCANNERS 33
  - RECOMMENDATIONS 34
- UNLICENSED RADIO SERVICES 36
  - CITIZENS BAND 36
  - FAMILY RADIO SERVICE 39
  - GENERAL MOBILE RADIO SERVICE 40
  - MULTI-USE RADIO SERVICE 42
  - HINTS FOR HT OPERATION 43
  - RECOMMENDATIONS 45

LICENSED RADIO SERVICES – AMATEUR RADIO 46
    OBJECTIONS TO AMATEUR RADIO LICENSING 48
CLASSES OF AMATEUR RADIO LICENSES 50
AMATEUR RADIO LICENSE TESTING 51
HOW TO GET STARTED IN AMATEUR RADIO 54
AMATEUR RADIO HF EQUIPMENT 57
    TRANSMITTER 57
    RECEIVER 58
    OTHER ACCESSORIES 58
ASSEMBLING AN AMATEUR RADIO HF STATION 62
    HOW MUCH DO YOU WANT TO SPEND? ... 62
    USED VS NEW 64
HF TRANSCEIVER CONTROLS 67
HF TRANSCEIVER REAR PANEL 73
ANATOMY OF AN AMATEUR RADIO HF CONTACT 76
AMATEUR RADIO RESOURCES 85
ASSEMBLING A VHF/UHF STATION 86
    EQUIPMENT FOR A VHF/UHF STATION ... 86
    CHOOSING THE EQUIPMENT 89
    HOW MUCH DO YOU WANT TO SPEND? ... 90
    RECOMMENDATIONS 91
REPEATERS 92
    RECOMMENDATIONS 95
ANATOMY OF A REPEATER CONTACT 96
MESH NETWORKING 97
OPERATING TIPS AND STRATEGIES 99
    COMMUNICATION PLANNING 99
    NETS 100
    PHONETICS 101
    HOW TO USE A MICROPHONE 101
    ENCRYPTION AND CIPHERS 102
ELECTROMAGNETIC PULSE (EMP) 104
GROUNDS, LIGHTNING AND EMP PROTECTION 106
    ELECTRICAL SAFETY GROUND 106
    LIGHTNING AND EMP GROUND 106

RF GROUND 108
ANTENNAS.. 113
    THE DIPOLE.. 113
    VERTICALS 115
    POLARIZATION 116
    NVIS ANTENNAS....................................... 117
    ANTENNA GAIN 118
    RECOMMENDATIONS 120
FEEDLINE 121
    RECOMMENDATIONS 122
MOBILE OPERATION................................... 123
LEARNING MORSE CODE............................ 127
TROUBLESHOOTING 131
    EQUIPMENT FAILURE 131
    I CAN'T HEAR ANYBODY, OR THEY CAN'T
    HEAR ME... 132
SUMMARY AND RECOMMENDATIONS.............. 134

APPENDIX A ANTENNA LENGTHS.................... 135
APPENDIX B 50,000 WATT AM STATIONS..... 137
APPENDIX C GMRS/FRS FREQUENCIES......... 144
APPENDIX D PREPPER RADIO FREQUENCIES 145
APPENDIX E BATTERY CAPACITY................... 146
APPENDIX F POWER CABLE 148
APPENDIX G POWER REQUIREMENTS........... 149

INDEX... 150

INTRODUCTION

I have taught many Amateur Radio classes and found that most folks coming into Amateur Radio today are preppers. That differs from when I started. We were electronics and science geeks.

Preppers want to communicate, not learn radio theory. However, without knowing some theory, you can fall prey to outrageous and outlandish advertising claims, misinformation, and rumor, any of which could be disastrous. This book will teach enough theory to understand radio's practical application.

I will show realistic expectations for various radio services, how to choose and assemble the equipment and fix things when they don't work. This book teaches The Easy Way to communicate when all else fails.

There are opinions on these pages, but I have avoided burdening the reader with qualifying words such as "usually, probably, and mostly." Radio is an art and a science, and there are exceptions and qualifiers for every rule. I discovered a long time ago, "If you ask five Hams a question, you will get seven different answers and maybe a fistfight."

I am not an electrical engineer, and you don't need to be one either. But I have been a Ham Radio operator long enough to offer insight into the world of communication.

Please send comments, corrections, and chastisements to K4ia@EasyWayHamBooks.com. This book prints on-demand, and corrections can appear within 24 hours.

COMMUNICATION FOR PREPPERS

I don't need to convince you of the need for disaster preparation. You've already considered the possibilities and bought this book. How serious are you? Easing the inconvenience of a few days satisfies some, while others prepare for years of travail.

This book assumes no cell service, no phones, no Internet, and no power. That drastic scenario is realistic enough for all levels of preparation.

Wherever one falls on the preparedness spectrum, communication is critical. There is comfort in knowing, and even bad news is valuable. Imagine being hunkered in a bunker and wondering, "Does New York City still exist?" Sobering thought. At the most basic level, you need to hear what is happening in the world around you. You need information to survive.

You need to transmit to contact others for aid and support. Don't limit the horizon to how loud you can yell. When the Internet and cell phones are down, radio is the only source of information and medium of communication.

THE IMPORTANCE OF PREPARATION

A prudent person foresees danger and takes precautions. The simpleton goes blindly on and suffers the consequences. Proverbs 22:3[1]

To be prepared, you must prepare. That means get ready now when there is no panic, no anxiety, and no immediate need. Once disaster strikes, it is too late. The time to gather equipment and plan communication solutions is now. The time to get ready is when you don't need to.

If you bought this book hoping to learn how to make a radio from a coconut, Gilligan's Island-style, or MacGyver a power supply with your Swiss-Army knife, sorry. That is only on TV.

You can turn most of my recommendations to fair use in good times, unlike pouches of freeze-dried sawdust or powdered water stored under the bed. Radio is something you can use and enjoy, as attested by 800,000 amateur radio operators in the US alone. Now is the time to learn how and where to listen for outside news and set up a communications network with your family or tribe. Use it and keep using it, so as not to forget. Regular on-air meetings keep everyone sharp and expose planning weaknesses. Build a communications network over time, if you must, but get started.

This book is about radio, and we start by exploring some radio theory to help understand what is to follow.

---

[1] New Living Translation

FREQUENCY

First is the concept of frequency. Electricity travels by direct current or alternating current. Direct current flows in only one direction, as from a battery, while alternating current reverses direction.

Frequency The number of times per second that an alternating current makes a complete cycle is the frequency. We measure frequency in cycles per second. The term "hertz" is a shortcut for "cycles per second."

The power in a wall socket cycles 60 times a second and is, therefore, 60 hertz. A radio signal may be anywhere from a few thousand hertz up to billions of hertz. WMAL, a local AM radio station, broadcasts on a frequency of 630,000 Hz or 630 kilohertz, abbreviated 630 kHz, or 630 thousand hertz. Citizens' Band radios operate around 27 MHz, 27 megahertz, or 27 million hertz (cycles per second).

The frequency of the radio wave determines the size of an antenna and how that wave will propagate (spread.) More on that later.

Electromagnetic energy A radio wave is electromagnetic energy. The reversing current causes an electromagnetic field to form around the antenna wire. That field radiates (spreads) and induces a current into the receiving antenna, even if that antenna is thousands of miles away. That is how radio works.

Radio is magic to me. A signal leaves the wire in my backyard and travels across town or to the other side of the world. To quote Samuel Morse's first Morse code message, "What hath God wrought?"

FREQUENCY

Electromagnetic spectrum We divide the electromagnetic spectrum into four categories:
- 300 kHz to 3 MHz (MF Medium Frequency)
- 3 to 30 MHz (HF High Frequency)
- 30 to 300 MHz (VHF Very High Frequency)
- 300 to 3000 MHz (UHF Ultra High Frequency)

Wavelength A radio wave travels through free space at the speed of light, approximately 300,000,000 meters per second. If we know the wave travels at 300 million meters per second and the frequency is 144 million cycles per second, we can calculate how far the wave will travel in one cycle. Wavelength is the name for the distance a radio wave travels during one complete cycle.

The magic formula is to divide 300 by the frequency in Megahertz (MHz) to get the wavelength in meters. Or divide 300 by the wavelength in meters to get the frequency in MHz.

The higher the frequency, the more often the wave reverses direction, and the less distance it can travel in a cycle. Therefore, the wavelength gets shorter as the frequency increases, and knowing the wavelength helps calculate the optimal length for an antenna.

Band A band is a group of frequencies, and referring to the band name is a shortcut. The AM radio band stretches from 550 kHz to 1600 kHz (Kilohertz), in the Medium Frequency range.

Bands are often associated with wavelength. The two-meter amateur band goes from 144 to 148 MHz.

We'll learn more later about the characteristics of different frequencies and why to pick one group of frequencies (band) over another to accomplish your communication goal.

MODES

The mode is the transmission type. A car radio has AM and FM modes, which are different means of modulating and receiving radio signals. Modulation describes combining speech with a radio frequency carrier. The RF carrier is the base frequency of the transmission. AM is amplitude modulation, and FM is frequency modulation.

FM MODE

FM mode (frequency modulation) communicates information by jiggling the transmission frequency, resulting in a relatively wide but clear high-fidelity signal. The high-frequency VHF and UHF bands use FM mode.

AM MODE

The AM Mode (amplitude modulation) communicates by jiggling the amplitude of the radio wave. The audio information varies the power of the wave, and the result is the center frequency called the carrier and a sideband on either side of the carrier, varying in amplitude in accord with the audio. Shortwave broadcast stations use AM. Most CB radios are AM mode.

SINGLE SIDEBAND MODE (SSB)

Single sideband (SSB) is a form of amplitude modulation (AM). Look at the diagram. The bottom waveform is an AM signal. To make SSB, filters in the transmitter strip away the carrier and one sideband, leaving only a single sideband. The remaining sideband could be on the upper

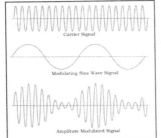

MODES

or lower side of the carrier. You will hear this referred to as upper sideband or lower sideband. By convention, frequencies above 14 MHz use the upper sideband (USB).

Single sideband sounds like Donald Duck until tuned properly. However, SSB is a superior voice mode because it requires less power and takes up less bandwidth (space) than AM or FM. Amateur Radio operators use SSB as the primary voice mode, and it is also available on some Citizen's Band radios.

PACKET MODE

Packet mode uses bursts of information (data). A modem creates and decodes the packets, checking for errors at the receiving end. If there are errors, the receiving station automatically requests a repeat. Packet is handy to transfer messages or lists of information.

CW MODE (MORSE CODE)

CW, which stands for continuous wave or carrier wave, is the most basic of digital modes. Turning the carrier on and off, makes the dots and dashes of Morse code.

Compared to other modes, CW has the narrowest bandwidth and the best chance of being understood under poor conditions.

Morse code is not required for any level of radio license, but Morse is still popular among Hams. It can provide a more discrete means of communication than voice because few people understand Morse. There is a chapter on learning Morse code later in the book.

POWER

A radio requires a source of power. If the mains power is out, we'll need a backup. How and how much requires understanding some electronic theory.

VOLTS, AMPERES, AND RESISTANCE

Think of voltage as the pressure in the line. Voltage is electromotive force measured in volts. A force of 120 volts is present in an electrical outlet, but the 120 volts do nothing until you connect to a load.

Amperes are a measure of the current or amount of electricity flowing through the circuit. Turn on a light, and current flows through the wire to the bulb. The voltage pushes the current through the load (light bulb).

Resistance, or load, also affects the flow, and with more load or resistance, less current flows.

Power, measured in watts, is the amount of work done by the pressure and flow.

There is a relation between all four:
More pressure = more current flow.
More resistance = less current flow.
More current or more pressure = more power.
Are these mathematically related? Georg[2] Ohm thought so and came up with his famous formula known as Ohm's Law.

Ohm's Law: Voltage = Amperes times Resistance.
The unit of resistance is ohms.

---

[2] That is not a typo. Georg has no "e."

POWER

E symbolizes voltage (Electromotive Force).
I is Amps or Amperage (flow).
R is Resistance.

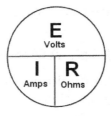

The formula for Ohm's Law is E=IR. Knowing two of the variables, you can calculate the third. The easy way is to use the magic circle and draw it as the Eagle flies over the Indian and the Rock.

Put a thumb on the value you are solving, and the formula is what's left. For example, if solving for R, cover up the R, and the answer is E/I, voltage divided by amperage. If you are solving for E, cover it up, and the answer is I x R, amps times ohms.

A practical use might be to calculate the voltage drop in a power cord. Longer or thinner power cords have more resistance, and the voltage drop is amperes times the resistance in the cord. Using a cable too thin or too long will cause an unacceptable voltage drop starving your radio or power tool. They won't perform as expected. The tables in Appendix F show recommended cable sizes for various loads.

POWER IN WATTS

Watts is a measure of electrical power or work. Calculating watts tells what power is required from a generator or battery.

The formula to calculate electrical power in a DC circuit is Power equals Voltage (E) multiplied by Current (I). It is easy as PIE, P=IE

POWER

Another magic circle helps us remember.

If we know a particular piece of equipment uses 100 watts at 120 volts, it draws about 100/120 or .8 amperes (amps). If it uses 100 watts at 12 volts, it will draw 100/12 or 8 amps.

The lower the voltage, the more amps are required to produce the same power.[3] Referring to Ohm's law, the more amps, the more voltage drop. Keep low-voltage lines, like portable solar panels short or use very stout wire.

GENERATORS

A fuel-powered generator is overkill to run a radio alone but can also power the rest of the household. The downside is that generators are noisy and distracting. They also create poisonous carbon monoxide gas and require a source of clean fuel.

Generators must be well-ventilated. Deadly carbon monoxide (CO) is a by-product of internal combustion engines. It is colorless and odorless, so there is no warning before you pass out. If you are sleeping, you won't wake up. Do not use generators indoors, in a garage, in enclosed areas, or near air intake vents. If someone acts confused or has cherry red lips or fingernails, suspect CO poisoning and immediately get them to fresh air, oxygen, and medical attention. Be safe and install a carbon monoxide detector in any sleeping areas.

---

[3] "The War of the Currents" has been the topic of many books and movies. Edison favored lower voltage direct current. Tesla advocated for high voltage alternating current that could be stepped down to usable voltages at the user level. Tesla was right.

POWER

Generator ratings refer to the peak and running power produced, measured in watts (volts times amps). The running power is called the sustained power, and it is less than the peak.

Don't consider just the running wattage. The starting load can be much higher because pumps, motors, and compressors can require three to four times the operating power to start. If a window air conditioner draws 1,200 watts running, it might need an additional burst of 3,600 watts to get going.

Appendix G is a list of common appliance power requirements.

Add the power consumption for all the equipment you intend to run simultaneously and choose a generator to handle that load. The total requirement might not be as much as you think because equipment like a refrigerator, freezer, and other appliances don't run continuously and would probably not all be running and drawing current simultaneously. You can coordinate the timing by unplugging equipment. A refrigerator or freezer will stay cold if you keep the door shut and might only require 15 minutes of power every couple of hours.

Whole-House Generators The most elegant (and expensive) solution is a whole-house generator. These turn on automatically when the power fails and energize selected circuits in the house. A safety breaker disconnects from the mains, so the power generated does not flow online.

Do not plug any generator into the mains electrical service unless there is a disconnect breaker. Your generator will overload and fail because it can't supply enough power for the entire neighborhood. You could also electrocute a utility worker.

POWER

Whole-house generators are expensive and, like all generators, need a source of clean and fresh fuel. Gas or diesel does not store well, and propane needs a large storage tank. Natural gas is an option if the gas lines are not interrupted, as might happen in an earthquake.

The supply line to my house was not large enough to handle a whole-house generator, and the gas company wanted $1,500 to install 10 feet of upgraded pipe. That, plus a $10,000 quote for the generator equipment and installation, turned me away from a whole-house solution.

Whole-House Battery Systems While not a "generator," whole-house battery systems serve the same purpose without the noise and hassle of fuel-powered systems. These charge from the power mains or solar sources. The Tesla Powerwall is a 13.5 kWh (kilowatt-hour) battery costing about $11,000, including installation. The average residential customer uses 29 kWh per day. A Powerwall would run an entire household for half a day or limited circuits for a week. Daisy-chain up to 10 Powerwalls for additional power.

Electric hybrid cars store energy, and the Teslas have a capacity of 50 to 100 kWh while a Toyota Prius hybrid's battery is 6 kWh. Wiring to access this power is a job for professionals.

For a complete discussion of solar possibilities, refer to "Energy choices for the Radio Amateur," by Bob Bruninga, available from ARRL.org or Amazon.[4] Bob was a big advocate of solar power and has lots of experience.

---

[4] https://www.arrl.org/shop/Energy-Choices-for-the-Radio-Amateur/

POWER

Construction Generators I refer to the gas or diesel-powered models on wheels as construction generators, and they provide 4,000 to 8,000 watts.

The problems with construction generators are that they:
- are big, heavy, loud, and hard to move
- run at a high, constant speed of 3,600 RPMs, no matter what the load, which consumes a lot of fuel
- may not produce clean, well-regulated power suitable for electronics.

Fuel consumption depends on the load. Heavier loads burn more. One gallon of gas or three pounds of propane lasts about three hours on a 50% load. Depending on the load, plan on 8 to 15 gallons of gas or 24+ pounds of propane for 24 hours' operation. You could stretch the number of days by only running the generator when needed. You will have to store and stabilize a lot of fuel.

The Uniform Fire Code prohibits storing over 5 gallons of gasoline in residential areas. Store gasoline or propane in an open place where fumes dissipate. Gas vapor is heavier than air, and explosive amounts can accumulate in a low spot.

The voltage and frequency from some construction generators are not well regulated, and do not produce a clean alternating current wave. While suitable for power tools and lights, electronics such as radios and computers might not work well.

Inverter Generators The inverter generator uses an electronic circuit to filter and regulate output. Inverters cost more but have several advantages over construction generators:
- much smaller, lighter, and quieter

POWER

- speed changes to match the load, so they consume less fuel
- power output is a well-regulated pure sine wave your electronics will appreciate.

My Honda 2000-watt inverter generator is light enough to carry and quiet enough to hold a conversation while standing next to it. Depending on the load, the fuel requirement is 2 to 6 gallons for 24 hours of operation.

While the 2000-watt rating seems low compared to construction generators, this may be all you need. It is certainly enough for radios. You can fudge with other equipment because things like refrigerators or freezers don't run continuously. If you keep the door shut, they will stay cold for hours. When we lost power during a hurricane, I kept going by cycling the appliances, including a small window air-conditioner. I kept the room cool, the beer cold, and the TV on for three days using less than 5 gallons of gas.

See the guidelines in Appendix F to size extension cords.

Inverter generators are more expensive than construction generators, but I would recommend them.

Stay away from the cheap 2-cycle models sold in discount houses. They are very noisy, both audibly and electrically, and have terrible voltage regulation.

POWER

BATTERIES

Batteries charged by solar panels require no fuel and are the best long-term power solution. How much solar panel and battery depend on what you plan to run. We will focus here on batteries to power radios.

Battery Types A supply of non-rechargeable alkaline batteries will store for years and run your low-power portable devices. An alkaline battery holder can back up a handheld device's rechargeables.

Rechargeables are better for more robust and sustained use, such as a handheld transceiver. The most popular rechargeable batteries include:
- Lithium-ion (Li-Ion)
- Lead-acid.

Li-Ion Li-Ion batteries, and their cousins, the LiPo (Lithium Polymer) and LiFePO4 (Lithium Iron Phosphate) recharge faster, are smaller and lighter, and hold more energy per pound than other battery types. Consider Li-Ion if you need to move around.

These provide 3.2 to 3.6 volts per cell. A battery comprises multiple cells connected to supply the rated voltage and amperage, 13.8 volts for our use. A special charging circuit balances the charge among the cells, and that circuit may be built into the battery pack. You must use a charger designed for Li-Ion batteries.

Li-Ion batteries require no maintenance, are one-third the weight and one-half the volume of lead-acid, and last in service up to seven times longer (2,000 charging cycles vs. 300). Li-Ion maintains its voltage through most of the discharge cycle and does not self-discharge, whereas other batteries experience voltage drop. A Li-Ion battery can discharge 80 to 90% and still maintain the rated voltage. A lead-acid battery

POWER

dies at 50%, and the voltage sags quickly while discharging. Hybrid cars and laptop computers use Li-Ion batteries.

Li-Ion batteries are the most expensive, but because of their long life and deep discharge cycle, they offer the greatest long-term value.

Lead Acid Lead-acid batteries at 12 to 14 volts (nominal) are the least expensive alternative energy source, but they are the heaviest choice. The sealed lead-acid design prevents spills. AGM (Absorbed Glass Mat) design is sealed and uses glass mats that cushion the lead plates for longer life.

Automobile starter batteries, designed to crank a car engine, produce a large amount of power in a quick burst. A lead-acid starter battery will lose voltage quickly once the charge goes below 90% of total capacity. That means only 10% of the battery capacity is usable at full voltage.

For sustained applications, use a deep-cycle trolling motor or golf cart battery. Their design produces power over a longer cycle, perhaps 50% of capacity. Take them lower, and you will permanently damage the battery. When idle, lead-acid batteries can lose 4% of their charge every 30 days. Li-Ion batteries only lose 1% in a year.

Lead-acid batteries contain acid and require careful handling and storing. Even so-called sealed batteries out gas acid vapor that will eat a holes in concrete floors or carpets.

Charging Use a solar panel to recharge a battery or charge by connecting it in parallel (positive to positive and negative to negative) with a vehicle's battery and idling the engine. When a disconnected lead-acid battery measures over 12.6 volts, stop charging. Use

POWER

a Li-Ion charge controller with Li-Ion batteries. A lead-acid charger doesn't have a high enough voltage to charge a Li-ion battery fully.

Consider a car as a source of energy in a power outage. A full tank of gas should run 72 hours and, charging a few hours a day, would last for weeks. Heed the warnings about carbon monoxide poisoning and don't do this in a garage.

Battery Dangers Charge controllers contain circuits to regulate the charge cycle. Make sure they match the battery type. Using a Li-Ion charger on a lead-acid battery would be a dangerous mistake because the Li-Ion charges at a higher voltage. A battery charged or discharged too quickly can overheat, give off flammable hydrogen gas and explode.

Another safety hazard is that shorting the terminals can cause burns, fire, or an explosion. Batteries can deliver a tremendous amount of current capable of melting a ring or tool. Take your ring or metal watchband off and keep tools clear when working around batteries. Seal battery terminals with insulating tape to prevent accidental contact.

Battery Capacity Batteries ratings are in amp-hours (Ah). "Amp" is an abbreviation for ampere, the amount of current flow in a circuit. One would think a battery rated at 4 amp-hours should produce 1 amp for 4 hours or 2 amps for 2 hours. It doesn't work out that way because a battery provides less power as it discharges, and the charge won't hold as long if discharged quickly.

A C20 rating is common. It means the amp-hour battery capacity is based on a 20-hour discharge rate, so a 4 amp-hour battery will supply 4/20 or .2 amps for 20 hours. That is enough for a handheld

POWER

transceiver and not much else. Draw more power, and the battery won't last.

Another measure of battery capacity is watt-hours (Wh), which is the amp-hour rating times the voltage. Either measure tells how much power the battery can deliver and for how long.

Unless adjusted, the rating assumes discharging a lead acid battery to 10.5 volts. Most radios will suffer once the voltage gets below 12 volts, so the radio will quit long before the battery. A Li-Ion battery will hold 13 volts for 90% of the discharge cycle and powers the radio longer.

A complete discharge would permanently damage any battery. To maximize battery life, a good rule of thumb is half the stated rating for lead-acid and 75% to 80% for Li-Ion. Use the amp-hour or watt-hour rating to plan and compare batteries, but don't expect to get the stated performance under actual conditions.

If the transceiver outputs 50 watts, 10 to 12 amps peak input is required at 13 volts. A 7 amp-hour lead-acid battery de-rated by 50% would only transmit 20 minutes.

The calculation gets complicated when applied to real life. A transmitter might need 10 amps peak when transmitting, but it isn't transmitting all the time, and some modes don't draw 100% when transmitting. FM draws 100%. Single sideband draws full power only on voice peaks. Average SSB conversation draws 20%.

The operational duty cycle is such that you are mainly listening and might transmit only 10% of the time. A typical radio could only need 2 amps while listening, so the 7 amp-hour battery in the example above would last much longer than 20 minutes. Further

POWER

conserve battery life by reducing transmitter power or listening with earphones rather than through a speaker.

How much battery do you need? The answer depends on how much power and for how long. You can do the math estimating duty cycles or accept recommendations from others who have. Bigger is always better.

Appendix E is a chart for LiPO4 batteries. It assumes a 20% transmit and 80% listening duty cycle. Lead-acid batteries cannot discharge as far as LiPO4s (50% vs. 80%), so the figures must be de-rated by at least 60% for lead-acid batteries.

As a general rule of thumb, to last through 12 hours of operation requires, at a minimum, one amp-hour of lead-acid capacity for each watt of transmitter output. That is the recommendation of the Radio Amateur Civil Radio Service (RACES), a standby radio service established by FEMA and the Federal Communications Commission. By this measure, a 50-watt transmitter would need a 50 amp-hour battery. RACES assumes doing a lot of transmitting during a 12-hour shift, and this standard is overkill for home use.

Another general answer is a deep-cycle lead-acid battery of at least 30 amp-hours capacity to listen mainly and transmit intermittently at 25 watts while running a computer or other accessories. That battery will weigh 25 pounds and cost about $100.

An equivalent Li-Ion battery at almost twice the cost will produce full output down to an 80% discharge. Replace the 30 amp-hour lead acid with an equivalent Li-Ion battery that is rated 20 amp-hours, weighs 5 pounds, and costs $170.

POWER

With either chemistry, monitor the voltage, so it does not drop below the battery's critical level. An inexpensive battery capacity monitor will show the battery's health.

Inverters A battery system outputs a nominal 12 to 14 volts DC. To convert that to 120-volt AC for household appliances, use a DC to AC inverter. The cheaper versions create alternating current by pulsing direct current, which may cause trouble operating radios and computers. Pay a little extra for a "pure-sine wave" inverter for best results.

Inverters can require substantial power from a battery. Charging a cell phone or laptop computer is trivial. The laptop might only draw 50 watts or 4 amps at 12 volts. (50/12=4)

However, if an appliance requires 400 watts at 120 volts, the inverter must draw 500 watts from the battery, allowing for inefficiencies in the inverter circuit. At 12 volts, 500 watts require 42 amperes from the battery (500/12=41.6). That is a lot to pull from a static battery alone, and the battery would discharge quickly. It will strand you if you rely on the same battery to start the car.

At higher power, attach the inverter to a car battery with the motor running. Attach the inverter directly to the battery because a cigarette lighter plug can't handle that much current.

Solar Power Extend the battery's discharge cycle or recharge a battery by attaching a solar panel to replace the energy used. If 10 amps come out of the battery, you need to put 10 back in to maintain the charge. A solar panel rated at 100 watts will output 5.5 amps at 17 volts (P=IE). The charge controller will reduce the 17 volts to 14 for charging. It will take 10/5.5 = almost 2 hours to recharge. The actual time

POWER

will be longer (maybe twice as long) because of inefficiencies in the charge controller circuits and less-than-ideal sun conditions. Judge solar panel needs accordingly.

Be wary of solar sellers' power capacity claims. Their figures assume clear days, bright sun, perfect alignment, a temperature below 75F, a spotless panel, and a bit of puffery. It will require more panels and more time to recharge than rated. A rule of thumb is to de-rate all claims by half. If the system produces more, be pleasantly surprised. It is better to be over-prepared than bitterly disappointed by dead batteries.

How to Extend Battery Charge The number-one way to extend a battery's charge is to reduce the radio's transmit power. FCC rules say to use the minimum power needed to maintain contact. If the contact is good at the transceiver's medium-power setting, don't use high. Try the low-power setting.

Half power requires half amperage, a considerable saving that conserves the battery and doubles the transmit time. It also reduces interference and lowers the chance of the signal being overheard by others because the signal does not travel as far.

POWER SUPPLIES

Solid-state transceivers require a nominal (about) 13.8 volts DC. That can come from a battery system, a car, or a power supply to convert your generator or house voltage (nominal 120 volts, AC).

Refresher: Volts measure pressure in the line. Your household electric outlet has 120 volts of pressure. Amperes or amps measure the current flow in the circuit; volts times amps = watts (power).

POWER

The current rating of a power supply is the number of output amps it can provide before the voltage sags or the fuse blows. For example, a "25-amp supply."

At 100 watts output, a typical transceiver will require 20 amps peak from the 13.8-volt source. Add a few amps for overhead, such as the other circuits in the radio, dial lights, etc. You can estimate the supply needs to be at least 23 amps.

Don't scrimp and get a 23-amp power supply for a 100-watt transceiver. It will be marginal and leave you no room for accessories, such as another radio. A 25-amp supply is adequate, but 35-amp is better[5]. The additional cost of the larger supply is not significant, and you won't be sorry. Meters also add little to the cost. I like a supply with volt and amp meters to see what is happening.

Linear Power Supplies There are two designs for AC power supplies. The so-called "linear" types use a large transformer to convert the 120 volt AC household current to the radio's 13.8 volts. Linear supplies are heavy and very rugged, and mine has been on almost continuously for 15 years without a glitch.

Switching Power Supplies Switcher type power supplies are much lighter and smaller. They convert the house mains 60 hertz AC to a higher frequency, allowing for smaller and lighter transformers to reduce the voltage. The frequency conversion can generate radio frequency interference (RFI), and switcher supplies have gotten a bad rap on that account.

Also, the switcher circuits are one more thing to fail. Check the reviews before buying and pay particular attention to complaints about RF hash. A supply

---

[5] A general rule of prepping is to always upsize.

POWER

designed to power radios has extra filtering and is less likely to cause issues. I have used both linears and switchers with no problems. Consider a switcher if you need to move the power supply often or carry it while traveling.

Keep all power supplies away from the transceiver to reduce the coupling of noise or hum into the signal. Setting them close together is asking for trouble. My supply is on the floor under the desk. It isn't something I adjust to, so I don't need access. I turn it on or off with my big toe, but mostly, it just stays "on."

Power Cable A new transceiver should come with a sufficiently stout power supply cable and the proper plug to fit the radio. A used transceiver may have neither. Don't scrimp with cheap speaker wire to power the rig.

Resistance in the wire causes the voltage to drop. If the transceiver draws twenty amps, a six-foot power cable made of number 16-gauge speaker wire will decrease the supply voltage by two volts at the radio. Low voltage starves the transmitter of the power it needs, and sagging power will distort the signal or shut down the radio completely.

A six-foot power cable is twelve feet of wire when considering both the positive and negative leads. Twelve feet of 10-gauge[6] wire will only drop about half a volt. Use 8 or 10-gauge wire designated as a power cord. With 12-gauge, keep the total length seven feet or shorter. See Appendix F.

If only a thin cable is available, run wires in parallel on both the positive and negative sides. Two wires on each side reduce the resistance and voltage drop by

---

[6] Lower gauge numbers mean thicker wire.

POWER

half. The proper supply cord will avoid many hard-to-diagnose headaches.

Watch the polarity! Attaching the power cord backward can severely damage a radio. Power supplies have positive and negative connectors. Power cords have one red and one black wire. Red is positive, and black is negative. Anderson Power Poles are a standard plug-in connector with one red and one black side, making it harder to reverse polarity.

Both wires should have fuses in a mobile installation. At home, at least fuse the red wire.

RECOMMENDATIONS

- For a bug-out source of radio power, a LiPO4 battery rated at least 20 Ah
- For a stay-at-home source, a lead-acid battery rated at least 30 Ah
- Recharge batteries with a 100 to 200-watt solar panel using a charge controller designed for the battery's chemistry
- Install a mobile radio in your car to use the car's battery and charging circuit
- A plug-in power supply for a base radio could be a switcher or linear. If you're moving around, use the lighter and smaller switcher design
- For additional transportable power, consider an inverter generator.

PROPAGATION

Propagation is the term used to describe the spread of radio signals. The most significant influence on HF (High Frequency or Shortwave, 3 – 30 MHz) is the ionosphere, comprising clouds and layers of charged particles above the earth. HF propagation changes with the seasons, sunspot cycles, and during the day or night.

UHF and VHF signals (30 MHz and above) do not bounce[7] off the ionosphere. UHF and VHF frequencies are line-of-sight and best suited for local communications.[8] FRS, GMRS, and two-meter Ham are on UHF/VHF. Another chapter covers these.

UHF and VHF frequencies can bounce off buildings and obstructions, producing multi-path distortion or echoes if they bounce off multiple surfaces. That would cause a fluttery or distorted signal. Move to get out of the echo chamber.

HF signals (3–30 MHz) can bounce off the ionosphere, but the various frequency bands react differently. During the day and periods of high solar activity, the general rule is that higher frequencies are open. The lower frequencies are open at night. The twenty-meter Amateur band (14 MHz) is in the middle and can be open to somewhere in the world 24 hours a day. Above and below twenty meters, the general rule is most noticeable.

To establish a long-distance two-way radio link, without help from the Internet, you need an Amateur Radio license to transmit on HF.

---

[7] The more scientific term is "refract."
[8] VHF signals can benefit from atmospheric conditions called tropospheric ducting but that is not predictable or reliable for our purposes.

PROPAGATION

Shortwave radio is a term used to describe international broadcasting on the HF bands. Shortwave is not nearly as active as before the Internet, but you can still hear many stations. A lot of shortwave broadcasting is news mixed with cultural, propaganda, or religious messages. Listening without transmitting does not require a license.

You can also listen to Hams, and they will probably comment on their local conditions.

Here is a guideline for Amateur and Shortwave radio bands and HF propagation:

Band in Meters	Band in MHz	Characteristics SW – Shortwave broadcasting AR – Amateur Radio
160 M	1.8 – 2	Amateur Radio (AR) Nighttime only.
120 M	2.3–2.495	SW Local in tropical regions. Time stations at 2.5 MHz.
90 M	3.2–3.4	SW Local in tropical regions with limited long-distance nighttime reception. Canadian time station CHU is on 3.33 MHz.
75-80 M	3.5 – 4.0	AR Local during the day and longer at night.
40 M	7 – 7.4	SW Mostly Eastern Hemisphere. Shared with the North American Amateur Radio Band.
60 M	4.75–5.06	SW and limited AR. Local in tropical regions. WWV time station on 5 MHz.

Prepper Communications

PROPAGATION

Band	Freq (MHz)	Description
49 M	5.8–6.2	SW Nighttime band. Poor for long distances during daytime.
41 M	7.2–7.45	SW Reasonably good nighttime reception especially from Europe. Shares part of the 40 M Amateur Radio band.
40 M	7.125–7.3	AR voice band. Several hundred miles daytime. Long distances at night.
31 M	9.4–9.9	Most used shortwave band. Good year-round nighttime and best daytime reception is in winter. WWV time station on 10 MHz.
25 M	11.6–12.1	SW Best during summer and just before and after sunset all year
22 M	13.57–13.87	SW Similar to the 19 M band. best in summer.
20 M	14–14.350	AR Long distance reception most of the day and night. Voice above 14.150
19 M	15.1–15.83	SW Day reception good, night reception variable; best during summer. Time station WWV on 15 MHz.
17 M	18.068–18.168	AR Best during daytime. Voice above 18.110
16 M	17.48–17.9	SW Day reception good. Night reception varies seasonally, with summer best.
15 M	18.9–19.02	SW Lightly used daytime band

PROPAGATION

15 M	21. – 21.450	AR Voice above 21.2. Trans-equatorial contacts are common during the day.
13 M	21.45– 21.85	SW Erratic daytime reception, with very little night reception. Requires good solar flux numbers.
12 M	24.890 – 24.990	AR Voice above 24.930. Requires good solar flux.
11 M	25.6– 26.1	SW but seldom used. Reception is poor in the low solar cycle, but potentially excellent with good solar flux.
11 M	26.965 – 27.405	Citizens Band. Mainly local contacts.
10 M	28.0 – 29.7	AR Voice above 28.3. Requires good solar flux numbers for distance reception.

WHERE DO YOU NEED TO COMMUNICATE?

Propagation on the HF bands may be too long for your needs. You might hear signals from Europe, but nothing closer. The signals are bouncing over your head because you are in the "skip zone."

Decide where you need to communicate and design accordingly.   Communicate locally with CB or GMRS, as described in the "Unlicensed Radio Services" chapter. Don't assume something will work without experimenting. A building or other obstruction might impede signals between the two stations even though the path is not long.

PROPAGATION

Set up VHF/UHF directional antennas as high as possible on both ends of the circuit, and they might reach 50 miles. Details are in later chapters.

If the goal is to communicate with the state capital 100 miles away, the twenty-meter Amateur band (14 MHz) is not the answer. Neither is CB or GMRS. I would be on 80 or 40 meters as a Ham using a Near Vertical Incidence Skywave Antenna (NVIS). Read more in the Antenna chapter.

The time to solve these dilemmas is now when there is no crisis, no anxiety, no pressure, and no one's life is at stake. Practice and experiment during the easy times to prepare for the tough ones.

Remember the seven Ps:
Proper
Prior
Planning
Prevents
Pretty
Poor
Performance.

RECEIVERS AND SCANNERS

You will want to know what is happening locally, nationally, and internationally. Radio supplies news, comfort, and entertainment.

The CONELRAD radio system used the AM radio band as a national emergency broadcast network during the Cold War. CONELRAD emblems marked the radio dials, and the public was to tune there in case of an emergency. Other stations went silent, so enemy bombers couldn't use the signals as target beacons.

Sadly, this system does not exist today. Our preparedness is worse. Few commercial radio stations have sustainable backup power, and many only broadcast an Internet feed. If there is no power or no Internet, local stations are off the air.

FM The FM radio frequencies are too high to refract and travel long distances. FM stations are strictly local.

Long-range AM Local stations may be off, but there is long-range activity on the AM radio band at night. That is because signals at those frequencies refract (reflect) off the ionosphere and can be heard at a distance. Distance listening only works at night because another ionospheric layer blocks these frequencies during the day. Tune around after sunset for long-distance AM-band signals. Appendix B is a list of 50,000-watt AM mega-stations.

A car's AM/FM radio might not be adequate. For more consistent coverage, get a better receiver and antenna, such as a general coverage shortwave receiver.

Shortwave Shortwave radio operates in the HF (high-frequency) bands between 3 and 30 MHz. At these frequencies, signals can bounce off the

RECEIVERS AND SCANNERS

ionosphere day or night, traveling hundreds or thousands of miles.

A "general-coverage" receiver hears the AM and shortwave bands. A radio that also receives SSB (Single Sideband) will pick up Amateur Radio operators. They can be a good source of information, particularly if tuned in to an emergency service net.[9] You can find a list of Amateur Radio nets on the Internet.

NOAA Some receivers also cover the National Oceanic and Atmospheric Administration (NOAA) weather alert frequencies on seven channels between 162.400 MHz to 162.550 MHz. These channels carry continuous 24/7 bulletins from the National Weather Service.

NOAA has Specific Area Message Encoding (SAME) for weather and non-weather emergency messages. If the radio is SAME enabled, key in the six-digit code for your area to hear local emergency alerts break into the regular weather reports.

RECOMMENDATIONS

- A quality AM/FM/Shortwave/NOAA radio
- One that is also SAME enabled is even better
- They come in all shapes, colors, and sizes
- Multiple power sources are helpful, and some will accept batteries, solar power, or a crank to charge an internal rechargeable battery
- Avoid toys and get a decent radio with a good-sized speaker and headphones as you may need to spend hours listening

---

[9] ARRL.org/ARRL-Net-Directory-Search

RECEIVERS AND SCANNERS

- Radios with frequency synthesizers and digital readouts are better (more precise and stable) than the dial models.
- Spend at least $100 and get a cheaper model to keep the kids entertained
- Become familiar with the shortwave bands to learn the productive frequencies.

Hints for operation:
- For the best shortwave reception, extend the antenna using an alligator clip and 10 to 20 feet of wire; extra length will increase the signal gathering ability; it might also be "too much" and overload the receiver; experiment
- Antennas are directional; move the antenna for the best reception; get it up high; try hanging it vertically
- Don't store the radio with batteries installed; they will die and leak, often ruining the radio; store batteries separately
- Play with the radio to familiarize yourself with the operation and learn the frequencies and stations you can reasonably expect to hear.

Scanners Scanners scan a group of frequencies and stop whenever there is a signal. Program the scanner to check a range of frequencies. Most HTs (handheld) and shortwave radios have scanning functions.

Dedicated police scanners were fun before police and emergency services started encrypting transmissions. Now, the scanner will stop on a signal, but you won't understand the scrambled transmission. Scanners are still sold, but you can't understand a scrambled transmission. Before spending money on a dedicated scanner, check one of the many online databases to see if it will be worthwhile in your area.[10]

---

[10] http://www.interceptradio.com/

UNLICENSED RADIO SERVICES

Listening is helpful, but talking requires a transmitter. National and international law regulates transmitting. Anyone can buy a radio and listen, but they may not transmit except as the laws allow. We'll start with simple services that do not require a license to operate.

The first question everyone asks is, "How far will it transmit?" The unlicensed radio services' answer is going to disappoint. Advertisers exaggerate wildly, and the promises could only happen under mountain-top to mountain-top conditions, sliding downhill both ways with the wind at your back.

CITIZENS BAND

CB was all the craze in the 1970s, and one might say it suffered from too much of a good thing. CB became popular and degenerated into a brawl. Rude behavior and offensive language have turned many people away. However, millions of CB radios are in place and carry action in an emergency.

CB radios operate on 40 channels[11] around 27 MHz, limited to 4 watts output of AM or 12 watts of SSB. For comparison, a night light is 5 watts, so CB is a low power resource. The FCC[12] tests radios entering the marketplace, and they are "type accepted." They cannot exceed those specifications, no matter what the guy in the CB shop tries to tell you. If the CB shop tinkered with the radio to exceed the limits, they have broken the law, and adjusting the radio alone won't produce much more power than the original design.

---

[11] Before 1977, CB was limited to 23 channels. If you see a used 23-channel rig for sale, it is more than 40 years old.

[12] The Federal Communications Commission regulates and polices communication services in the United States.

UNLICENSED RADIO SERVICES

You can't exceed the rating of the final amplifier transistor. If you do, it will burn out. Trying to get more power is like shoving ten pounds of potatoes in a five-pound bag.

You will see CB amplifiers for sale, commonly referred to as "kickers," "after-burners," "linears," or "pills." Amplifiers and souped-up radios are illegal and rarely meet spectral purity and filtering requirements. They can generate interference, and distortion.

Maybe you don't care if the radio is legal or not. You only intend to use it where an illegal radio will be the least of anyone's concerns. I get it, but a radio that has been tampered with or attached to an amplifier is probably out of specifications. That means a lousy-sounding signal that attracts attention, unsafe operation, or premature component failure.

The joke is on whoever pays money to get the extra power of a souped-up CB radio or amplifier. Line of sight, the curvature of the earth, intervening buildings, hills, and other obstructions limit the useable distance. Increasing power won't make much difference, if any. It will only make you louder to stations who might hear you, anyway.

The other consideration is that one needs to take the radio out of the box and use it to become familiar with the operation and develop a realistic assessment. You won't do that if worried about a visit from an FCC field agent. A lousy signal draws attention.

CB radios operate off 11 to 13.8 volts, which is the nominal voltage of a car battery. CBs consume little power and can be plugged into a cigarette lighter. Setting up at home requires a power supply to convert the house voltage to 13.8 volts. If there is no house power, use a battery. The radio doesn't need a large

UNLICENSED RADIO SERVICES

power supply or battery because CB is low power, drawing less than 2 amps when transmitting.

The most common CB antenna is a vertical single stick. Full-size is 102 inches long.[13] Shorter antennas have a built-in coil that makes them longer electrically. Such tricks come at a price, and the short antennas are not as efficient as full-size ones. Shortened CB antennas often come with a magnetic mount to stick on the car.

Communication is line-of-sight unless the wave bounces off the ionosphere. That can happen and support long-distance contacts on 27 MHz under the right conditions, but those conditions are rare and unpredictable. Therefore, the reasonable range of a CB radio tops out at 2 to 5 miles, more if in the clear or mountain-top to mountain-top. Hills and valleys will significantly reduce the range.

A decent CB radio is $100 and up, plus an antenna ($15 to $50). New CBs can get crazy expensive with fancy lights and seldom-needed features. Used CBs are cheap and plentiful. Pawnshops are full of them (probably stolen).

Hints for CB operation:
- Channel 3 is the prepper channel
- Channel 19 is the calling channel where most conversation starts; switch to another channel once you make contact
- Channel 9 is the emergency channel, often monitored by police; stay off except to report an emergency
- Mount your antenna high for maximum coverage
- Use SSB mode for maximum talk power
- Get power from a cigarette lighter socket.

---

[13] One-quarter wavelength. See the chapter on antennas.

UNLICENSED RADIO SERVICES

FAMILY RADIO SERVICE

HTs are handheld devices (Handie Talkies or Walkie Talkies), and several radio services use this format. There are HTs for Citizens Band, but they are more common on the higher frequencies like Ham, Family Radio Service (FRS), and General Mobile Radio Service (GMRS). That is because an effective antenna at CB frequencies (102 inches) is too long to fit on an HT. An antenna for FRS or GMRS frequencies is only 6 inches long.

FRS radios are limited to 2 watts output and operate in FM mode on set channels at frequencies between 462 and 467 MHz. Channels 8-14 are limited to ½ watt. FRS radios must have permanently attached antennas. There is no substituting for something better. FRS radio's range is minimal.

Hints for FRS operation:
- Channel 1 is designated the calling frequency; make contact there but move to another channel to continue the conversation
- Channel 3 is the emergency channel
- Agree on privacy tone codes for your group to reduce interference
- Always hold your HT vertically, so everyone maintains vertical polarization (See "polarization" in the Antenna chapter).

FRS radios are kid's toys. The rules purposely cripple them to restrict coverage. They don't have the range or flexibility to do much more than communicate to the backyard or serve as an intercom. Because they are cheap, there are many in use, and interference makes FRS unusable in crowded conditions.

UNLICENSED RADIO SERVICES

GENERAL MOBILE RADIO SERVICE

General Mobile Radio Service (GMRS) requires a license, but I include it in the unlicensed services because the license is only a registration; there is no test. One license covers the whole family and costs $70 for ten years.

Apply for a license online at www.fcc.gov/uls. First, register to establish an account with the FCC. That nets you an FRN number,[14] which is your identification for all future business with the FCC. If you are reluctant to let the authorities know where you are, use a PO Box address. The FCC uses electronic filings and requires an email address.

Choose a password and proceed to the application page for the GMRS license. Within 24 hours, the FCC emails with a link to the license. My link wouldn't open, so I logged on to the FCC.gov.uls website and, by searching my FRN, found my call sign, WRBX217.

You must identify by announcing your call sign in English (or Morse Code) at the end of a series of transmissions and every 15 minutes during a conversation. The full text of the GMRS regulations is in Subpart E, Part 95, Title 47 of the Code of Federal Regulations. Note this section on prohibited uses:

§95.1733 Prohibited GMRS uses.
    (a) In addition to the prohibited uses outlined in §95.333 of this chapter, GMRS stations must not communicate:
    (1) Messages in connection with any activity which is against Federal, State, or local law;
    (2) False or deceptive messages;
    (3) Coded messages or messages with hidden meanings ("10 codes" are permissible);

---

[14] FCC Rgistration Number

UNLICENSED RADIO SERVICES

(4) Music, whistling, sound effects or material to amuse or entertain;

(5) Advertisements or offers for the sale of goods or services;

(6) Advertisements for a political candidate or political campaign (messages about the campaign business may be communicated);

(7) International distress signals, such as the word "Mayday" (except when on a ship, aircraft or other vehicle in immediate danger to ask for help);

(8) Messages which are both conveyed by a wireline control link and transmitted by a GMRS station;

(9) Messages (except emergency messages) to any station in the Amateur Radio Service, to any unauthorized station, or to any foreign station;

(10) Continuous or uninterrupted transmissions, except for communications involving the immediate safety of life or property; and

(11) Messages for public address systems.

(12) The provision of §95.333 apply, however, if the licensee is a corporation and the license so indicates, it may use its GMRS system to furnish non-profit radio communication service to its parent corporation, to another subsidiary of the same parent, or to its own subsidiary.

(b) GMRS stations must not be used for one-way communications other than those listed in §95.1731(b). Initial transmissions to establish two-way communications and data transmissions listed in §95.1731(d) are not considered to be one-way communications for the purposes of this section.

GMRS shares a few frequencies with FRS but has a maximum power of 50 watts on some channels vs. 2 watts for FRS. Privacy tones are available, and GMRS radios can use an external antenna. Fifty watts is quite a bit of power but would only be available from a base/mobile station and not an HT.

The practical limit for an HT is about 8 watts. More than that needs a large enclosure and heat sink to dissipate power lost in the amplifier circuit. (No

UNLICENSED RADIO SERVICES

amplifier is 100% efficient). The breakdown of frequencies and power is in Appendix C.

GMRS is not allowed in Canada or north of "Line A." Line A is an imaginary line along the United States' northern border, including Detroit, Buffalo, Cleveland, and Seattle. There is also a "Line C" along the Alaska/Canada border.[15]

GMRS is limited to line of sight, as there is no reflection off the ionosphere at these frequencies. With a clear sightline, meaning nothing between the two stations, expect 10 to 30 miles coverage at 50 watts.[16]

GMRS allows eight channels for repeaters, and, if available, they will expand the coverage area. See the chapter on Repeaters. There are very few GMRS repeaters compared to ham.

A GMRS package with a 15-watt base/mobile station, magnetic mount antenna, and 2 Handhelds costs about $200.   Handhelds alone are $30 to $50 each. You can find them on Amazon or in Walmart.

MULTI-USE RADIO SERVICE

Another service in the "private, two-way, short-distance voice or data" category is Multi-Use Radio Service. MURS is very similar to FRS and GMRS but does not require a license or registration.

MURS is also available for commercial users, and store and warehouse workers often carry the radios.

---

[15] https://www.fcc.gov/engineering-technology/frequency-coordination-canada-below-470-MHz#LineA

[16] Your mileage may vary. All distance figures depend on surroundings and the height of antennas.

UNLICENSED RADIO SERVICES

MURS operates on five channels, around 151 MHz. This frequency is too high to skip off the ionosphere, so it is line-of-sight.

Two watts output is the power limit, and the antenna cannot be over 60 feet above the ground. On the ground, don't expect any better coverage than FRS (2-5 miles in the clear). With high antennas at both ends of the circuit, perhaps up to 10 miles.

MURS HTs cost about $25 to 30 apiece and are available on Amazon.

HINTS FOR HT OPERATION

- Use fresh or freshly charged batteries
- Low voltage batteries will drastically cut power and cause distortion
- Don't store the radio with batteries installed; they will die and leak, often ruining the radio
- Store batteries separately.
- Don't let rechargeable batteries sit idle; some varieties can self-discharge and no longer be chargeable
- Get up high; the higher the antenna, the further the radio can see; height extends the radio horizon
- If the signal was good, but now it is weak or distorted, try moving; you may be in a radio shadow or receiving the signal on multiple paths as it bounces off objects; moving just a few feet can make all the difference
- The antennas shipped with HTs are shortened versions of what physics would require;[17] improve performance by substituting a full-sized antenna if the radio will allow it

---

[17] See the chapter on Antennas. A full-wavelength in meters is 300/frequency in MHz. Take one-quarter of that for the length of a full-size quarter-wave vertical in meters. 1 meter = 39.4 inches.

UNLICENSED RADIO SERVICES

- GMRS and CB allow for directional antennas
- Avoid cross-polarization; antennas at both ends of the circuit should point the same way (up and down or horizontal)
- Elevate the antenna at least 15 feet; a painter's pole would work.

The chief advantage of unlicensed services is many users are already in place, and equipment is inexpensive. These radios are suitable for chatting around town to hear what others are experiencing. Unlicensed service may be all you need for a cross-town link.

COMPARISON OF RADIO SERVICES

Service	FRS	GMRS	MUR	CB	Ham
Channels	22 at 462 to 467 MHz	22 FRS + 8 Repeater at 462 to 467 MHz	5 around 151 MHz	40 Around 27 MHz	Unlimited within Ham bands
Power	Channel 1-7 2W Channel 8-14 .5W	Channel 1-7 5W Channel 8-14 .5W Channel 15 – 22 50W	2W	4W AM 12W SSB	200 - 1500W
Antenna	Fixed	Separate allowed	Max height 60'	Separate allowed	Separate allowed
Line of sight	Yes	Yes, but extended coverage with repeaters	Yes	Yes, with occasional ionospheric skip	VHF/UHF, yes, but many repeaters. HF is not line of sight, and distance is unlimited

UNLICENSED RADIO SERVICES

RECOMMENDATIONS

- A used mobile CB and magnetic mount antenna on the car are inexpensive and provide a mobile communication post for local contacts
- CBs are everywhere and can inform of local conditions
- A GMRS radio should be part of your kit as it runs higher power, can access repeaters if available, and communicate with FRS radio users
- GMRS will not be as crowded as CB; GMRS radios are not as common, and communication will be more discrete[18]
- GMRS and CB radios adapt well to car installation.

---

[18] Accomplish stealth communications with equipment and modes that are not in common use. Codes and ciphers are illegal.

LICENSED RADIO SERVICES – AMATEUR RADIO

Amateur Radio is a licensed service for individuals as opposed to broadcasting, which is to the public. As a youngster, there was little entertainment. Three channels of black-and-white TV don't occupy you for long. I got a shortwave receiver one Christmas and listened to international broadcast stations. Tuning across the bands, I heard people talking to each other. That was how I discovered Amateur Radio.

Immediately, I knew that was something I wanted to do. Listening to distant stations, fading in and out of the static, sometimes with distinctive polar flutter, but always with a sense of magic, planted a seed. As a Ham, I could graduate from a passive listener to a participant.

K4IA as WA4TUF in 1967

LICENSED RADIO SERVICES

Nobody knows where the term "Ham Radio" came from, but you'll often hear it describe the Amateur Radio Service. I suspect people use "Ham" because they have a hard time spelling amateur. Just kidding.

There are over 800,000 Amateur Radio operators in the United States alone, with over 30,000 new licensees every year and about 3 million Hams worldwide. They are knowledgeable radio operators ready to bring their equipment and expertise to public service or emergency uses. All this expertise and equipment didn't cost the US government a dime.

Amateur Radio's advantages The Amateur Radio service has many advantages over unlicensed services. Here are some critical distinctions:
- Amateur Radio has a wide variety of operating frequencies and modes to adjust for propagation conditions
- Hams have privileges on the HF bands, not limited to line-of-sight communication
- The maximum transmitter power allowed for Amateur Radio is 1,500 watts
- Amateur Radio has much more sophisticated and capable equipment, including cutting-edge computer integration and digital modes
- Amateur Radio has established operating protocols, message nets, and infrastructure, unlike the chaos you hear on CB
- Amateurs get educated and must study to pass a test for their license
- Amateur-maintained UHF and VHF repeaters are available to extend the range of HTs and mobile stations; Amateur repeaters are everywhere.

OBJECTIONS TO AMATEUR RADIO LICENSING

The primary objections are: "It costs too much to build a station," "I don't need a license to push a button and talk," and "It is too hard, and I can't learn Morse code." Let's deal with these.

It costs too much. There is no need to spend thousands of dollars. Amateur Radio isn't cheap, but one can get on HF for the cost of a video gaming system or smartphone. Handhelds cost less than $50. Amateur Radio costs very little compared to SCUBA diving, golf, or a bass boat.

A prepper might have well over $1,000 invested in freeze-dried gruel he will never eat, survival gear he will never use, and ammunition he will never shoot. Amateur Radio equipment won't sit idle under the guest-room bed and can provide years of enjoyment even if the worst never happens.

Take comfort in the fact Amateur gear holds value and consider the annual cost. If you choose to go all-out with new equipment and sell in five years, a $1,500 transceiver might fetch 75% of the new price. That would be a discount of $375 for five years of service or $75 a year. That won't buy a weekend of golf.

There are active markets in used Ham gear on eBay, eHam.net, and QRZ.com.

Used equipment depreciates even more slowly. A $500 used transceiver might reasonably expect to sell for very close to that in a few years. Say it sold for $450 after two years; the cost was only $25 a year.

We'll discuss specific equipment in another chapter.

OBJECTIONS TO AMATEUR RADIO LICENSING

I don't need a license. Anyone can buy a radio and mash a button to talk, but will they know what to do? Can they communicate effectively? The path to being prepared is preparation. A license will teach how to operate the equipment effectively and what frequencies, antennas, and times are best to communicate.

Once licensed, time spent on the air is fun and good practice for any level of disaster. Most radio clubs offer training and assist in public service events as practice runs for emergency communication incidents.

Sure, in a genuine crisis, the authorities may have too much on their plate to bother enforcing the Communications Act, but that is not the point. If you don't know how to operate the equipment, it is just a brick. Write your message on a piece of paper, wrap the paper around the transceiver, and throw it. That is how far you will get.

It is too hard, and I can't learn Morse code.
Morse code is no longer a requirement for any class of Amateur license.

And, no, the tests are not too hard. I have taught classes and taken people who have no math or science background through the material. They all pass the tests. Eight and eighty-year-olds have done it. You can too.

We'll talk about the tests in another chapter. You will want to get my EasyWayHamBooks license manuals.[19] Start with "Pass Your Amateur Radio Technician Class Test – The Easy Way."

---

[19] EasyWayHamBooks.com, Barnes & Noble and Amazon

CLASSES OF AMATEUR RADIO LICENSES

The Federal Communications Commission regulates airwaves in the United States, and one must earn a license from the FCC to transmit on Amateur Radio.

There are three Amateur license classes: Technician, General, and Amateur Extra, each conferring more privileges. The privileges break down roughly as:

Technician
- UHF and VHF Amateur bands above 50 MHz with some limited Amateur HF privileges
- Technicians are limited to 200 watts output on HF
- Mainly for local communications.

General
- All Technician privileges plus most Amateur HF frequencies
- A portion of each HF band is reserved for Extras
- 1,500 watts maximum output power
- Local and worldwide coverage possible.

Amateur Extra
- All frequencies allocated to the Amateur service
- 1,500 watts maximum output power
- Local and worldwide coverage possible.

AMATEUR RADIO LICENSE TESTING

The entry-level Technician test is not hard. Don't let the material intimidate you. There are a few multiplication and division questions, but no algebra, calculus, or trigonometry. There is no Morse code required for any class of Amateur license. Morse code is still very much alive and well on the amateur bands, and there are many reasons to learn it in the future. Don't worry about it for now. The General test is a little more challenging, but not much.

The Technician and General tests are each 35 questions. The questions and the answers are word-for-word selected from a published question pool. We know what will be on the test and can study with the confidence you are focused on the right material. The Technician pool has 424 questions[20], and the General has 462.

The pool is large, but only about one in a dozen questions will show up on your exam, one from each test subject. Many questions ask for the same information in a slightly different format, so there aren't 424 or 462 unique concepts. You can miss 9 of the 35 and still pass. You can miss every math question and pass.

Do you know what they call the medical student who graduates last in his class? "Doctor" – the same as the guy who graduated first in the class. My point is: pass, and no one will know the difference. To pass, answer 26 out of 35 correctly.

The questions are multiple choice, so you don't have to know the answer; just recognize it. That is a tremendous advantage for the test-taker.

---

[20] The new question pool after July 1, 2022 has 412 questions.

LICENSE TESTING

The best way to study for a multiple-choice exam is to concentrate on the correct answers and find a memorable word or phrase. On test day, the correct answer should jump right out. I use that teaching method in my EasyWayHamBooks license series.

There are classes listed online,[21] or check with a local Amateur Radio club. Class attendance is not required, but many find classes helpful.

Volunteer Examiners, organized under the authority of the FCC, administer Amateur Radio testing. VEs are local Hams who help you get your license. In most areas, a VE group is testing at least once a month. There are 14 accredited VE organizations, but ARRL and W5YI are the biggest. Search here[22] for a test site or talk to the local Ham club. Online testing is also available.

On test day, bring a picture ID, your FRN number,[23] a pen and pencil, and about $15. Ask ahead to get the exact amount and see if they prefer cash or check. The modest exam fee reimburses the VE's costs — no one profits except you. The FCC has imposed a $35 license fee, for new licenses. This is paid directly to the FCC after you pass the test.

---

[21] http://www.arrl.org/find-an-amateur-radio-license-class
[22] http://www.arrl.org/find-an-amateur-radio-license-exam-session
[23] Obtain a Federal Registration Number (FRN) from the FCC. The FRN is used only for your communications with the FCC. This is the same number you got when registering your GMRS license.
https://www.fcc.gov/wireless/support/universal-licensing-system-uls-resources/getting-fcc-registration-number-frn

LICENSE TESTING

If you bring a calculator, clear the memories. Turn off your cell phone and don't look at it during the exam.

There is a simple application (Form 605). Save a few minutes by downloading the form from the Federal Communications Commission (FCC) website and filling it out ahead of time.[24] If a site charges for the form, get off and go to FCC.gov, the official FCC website, where the form is free.

The Volunteer Examiner (VE) team will select a test booklet. You also get a multiple-choice, fill-in-the-circle answer sheet. The answer sheet has room for 50 questions, but Technicians and Generals should stop at 35. There is no time limit, but most finish in 30 minutes or less.

The VE team will grade the exam and tell you a pass/fail result. Some teams will tell the score but don't ask them to go over the questions or say what you missed. They don't know and don't have time to look.

After passing, there is more paperwork, resulting in a CSCE, a Certificate of Successful Completion of Examination. The team processes Form 605 and CSCE, and your call sign appears in the FCC database in a few days. Once the call sign appears, you may operate. Congratulations.

Print a copy of your license from the FCC website. The license is valid for ten years and can renew without testing again.

---

[24] https://www.fcc.gov/fcc-form-605

HOW TO GET STARTED IN AMATEUR RADIO

Here are a couple of tips to get started in Amateur Radio:

- Join your local Amateur Radio club.
- Find an Elmer.
- Attend Hamfests.
- Take part in Field Day.
- Get on the air.

Join Your Local Amateur Radio Club Lists of local radio organizations are at ARRL.org/find-a-club.

There are clubs everywhere, and most host a repeater. Listen to the repeater to learn more about the group. Attend a meeting. Clubs are eager to receive new members and encourage new Hams. Don't be  intimidated if the discussion is over your head; you just need time to catch up.

Meet some folks. Join in on the repeater and some club activities and see if the group is a fit. Some emphasize public service, providing communications for marathons, bike races, and other events. Other clubs may be interested in contesting or just eating breakfast. If you don't feel the love, try another organization.

Find An Elmer The local club will provide a chance to rub elbows with experienced operators and learn from them. Joining a local club is a great way to fulfill the second bit of advice: find an Elmer. An Elmer is a mentor or coach, not a formal commitment, and you wouldn't ask, "Will you be my Elmer?" You're not "going steady" or anything like that. Look for someone who seems friendly, knowledgeable, and willing to answer questions.

HOW TO GET STARTED

Your Elmer is a "go-to" person when you are stumped or don't understand the jargon. There might be more than one Elmer. One is antennas' guru, another is a contesting fanatic, and another understands computers. Most Hams will share their expertise, so don't be afraid to ask questions.

There are many Facebook and Email groups and YouTube videos that are computer-age versions of Elmers.

Attend Hamfests A Hamfest is a Ham Radio-oriented gathering to swap/buy/sell, attend forums, meet for an eyeball QSO[25], demo equipment, and have a good time. It is a convention that mixes a flea market, museum, vendors, and camaraderie. They are a source of reasonably priced used equipment and connectors and cables that aren't economical to ship if ordered online.

Hamfests can be elaborate multi-day affairs attended by tens of thousands or local and low-key tailgate "junk-in-the-trunk" gatherings. The big hamfests include the annual Dayton Hamvention in May, the Visalia California DX Convention in April, and the Orlando Hamcation in February. Lists of upcoming events are at ARRL.org/hamfests-and-conventions-calendar.

Participate In Field Day Field Day is a long-standing Amateur Radio tradition. The last full weekend in June, Hams take to the great outdoors and operate using emergency power.[26] The goal is to set up a station and contact as many other stations as possible during the 24-hour activity.

---

[25] QSO is a conversation. "Eyeball" means meet face-to-face.
[26] There are other operating categories as well but this is the main one.

HOW TO GET STARTED

This event combines emergency-preparedness training, contesting, picnic, and camp out. Your local club probably takes part in Field Day. Field Day is an opportunity to see antennas erected, stations set up, and operators operating. You can take part as a helper or operator.

The rules allow for a Get-On-The-Air (GOTA) station restricted to new or "generally inactive" Hams. An unlicensed person may operate under supervision. Ask about the GOTA station and take advantage of the opportunity. Don't miss Field Day. It is an exciting time.

Get On The Air The last piece of advice is, "Get on the air." The way to get started is to start. No matter how modest your station is, you can make contacts. Operating tips and techniques are coming up in another chapter.

Ask a local Ham to visit his shack if you don't have a station. Most Hams are proud to show off their equipment and introduce you to the airwaves.

AMATEUR RADIO HF EQUIPMENT

The typical configuration is a transceiver comprising a transmitter and receiver in one box.

TRANSMITTER

When selecting a transceiver, the transmitter section is not nearly as important as the receiver, which will dictate your choice.

Fortunately, this is not much of a dilemma. Most HF transceivers used by Hams deliver 50 to 100 watts, and watts are watts no matter where they originate.

Rules limit GMRS radios to 50 watts. In most instances, that is sufficient power and will provide the intended communications over reasonable paths.

Just accept whatever the transmitter portion offers if you are on a budget. The basic radios will suffice. Concentrate on buying the best receiver.

Helpful but non-critical, features include:
- Built-in antenna tuner (not used on VHF and UHF)
- Built-in power supply (rare with modern equipment)
- Additional band coverage (6 meters and VHF/UHF)
- Built-in speech processor to tailor your audio
- Adjustable power settings, even if they are only "high" and "low."

AMATEUR RADIO HF EQUIPMENT

RECEIVER

When comparing transceivers, concentrate attention on the receiver. That is the most critical component.

Sensitivity is not the issue as they are all about the same and will receive at the natural background noise floor. What sets an excellent receiver apart from a mediocre one is selectivity and particularly selectivity to resist intermodulation byproducts from nearby strong signals. Intermodulation Distortion (IMD) is an unwanted mixing of signals and sounds like growling or distortion. Cheap GMRS gear can suffer from IMD.

A disadvantage of GMRS equipment is that it does not support the wide variety of interference-fighting tools found in Ham equipment. Also, the frequencies are on set channels, while a Ham radio can change frequencies to avoid interference.

Modern Ham radios use Digital Signal Processing. Digital Signal Processing converts the signal to numbers (1's and 0's), looks for patterns, and then manipulates the numbers to cancel out interfering signals and noise. The math is staggering, but microprocessors in the radio make it possible.

OTHER ACCESSORIES

The station can be much more than a transceiver and antenna (more on them later). There are dozens of accessories, but you do not need all these gadgets.

Antenna Analyzer Measures how well the antenna system matches the radio. An analyzer differs from an SWR[27] meter. An SWR meter uses the transmitter's

---

[27] Standing Wave Ratio is a way of expressing how well the antenna system matches the radio. Ideal is 1:1 but anything less than 2:1 is acceptable and even 3:1 is usable.

AMATEUR RADIO HF EQUIPMENT

power, only reads at one frequency at a time, and only measures forward and reflected power.

An analyzer doesn't need a transmitter, as it generates a low-power signal and displays a range of measurements over different frequencies. The analyzer will also reveal a defective connector or cable.

Anderson Power Poles are standard plug-in power connectors for attaching a power supply and equipment.

Baluns and line isolators may be the cure for RFI (Radio Frequency Interference) in an HF station. See the Chapter on "Grounds and Lightning Protection."

DC Power Distribution Panel A distribution panel with multiple connectors makes it easier to connect several pieces of equipment to the DC power supply. Some use Anderson Power Poles for convenience.

Dummy Load Test or tune-up into a dummy load, so the signal does not go out over the air. It is a large resistor, sometimes air-cooled and sometimes in a bucket of oil.

Filters There are many types of filters serving different purposes. AC line filters reduce voltage spikes and block noise on the AC power line. Audio filters can use tuned circuits or digital signal processing to reduce noise or narrow the audio passband to reduce adjacent interference.

RF filters suppress RF energy outside their design frequency. An RF filter might help avoid overload from a nearby powerful AM radio station.

Headphones Don't spend a fortune on high-fidelity studio equipment because deep bass and tinkling

AMATEUR RADIO HF EQUIPMENT

highs don't matter. The audio of voice communications is narrow and mid-range. The most important specification for headphones is that they must be comfortable and not crush your head in a vise.

There are three basic headphone designs. Over-the-ear models cover the ear entirely and block out external noise. They can become uncomfortable if worn for too long but are helpful in a noisy environment.

On-the-ear designs sit on top of the ear and provide little noise-blocking but are more comfortable. I prefer on-the-ear unless I am in a noisy environment.

Earbuds fit in the ear, are better noise blockers than on-the-ear designs, and are more comfortable than over-the-ear. I like earbuds for long sessions.

I have all three designs for different conditions and to provide some variety during long listening sessions.

Microphone Most new transceivers come with a hand-held microphone. It will be adequate. Headsets and desk microphones with a foot switch are easier to operate, freeing your hands for something else.

Multi-Meter This handy device, sometimes called a volt/ohm meter or VOM, measures volts, amps, and resistance. Use it to check voltage or for a broken wire. The cheap versions (under $10) are not laboratory-grade but are sufficient for most of our needs. This is a must-have accessory.

Speaker The transceiver's built-in speaker is not ideal. It fires upwards instead of at you. It is also small. Wide fidelity is not essential, and you might find that built-in speaker sufficient. An external speaker will improve your reception. Don't pay a premium price to get a speaker that matches your rig.

AMATEUR RADIO HF EQUIPMENT

When they say it matches, it only means they painted it the same color.

Speech Processor tailors audio to increase readability by adjusting the frequency response to emphasize the mid and mid-high frequencies that carry the most information. VHF/UHF FM doesn't use speech processing.

SWR Meter The SWR meter goes between your transmitter and the antenna to measure forward and reflected power while transmitting. Most are calibrated for HF frequencies. To use an SWR meter with a GMRS radio, get one that covers VHF/UHF frequencies.

The best accessory is an assistant, as shown on this QSL card from the Canary Islands. Hams exchange postcards, called QSL cards, as written proof of their contacts.

ASSEMBLING AN AMATEUR RADIO HF STATION

The Amateur Radio HF station is more complicated than the typical VHF/UHF station. There are more frequencies and modes with Ham, and the antennas are much larger.

HOW MUCH DO YOU WANT TO SPEND?

Moderate Used A moderate level used Ham station would be an all-mode (AM, SSB, CW, Digital) transceiver five to fifteen years old. It will cover the basic HF Ham bands and may receive in the shortwave bands (called "general coverage"). This equipment is modern enough to give plenty of service, but won't be on the cutting edge of technology. It would not have been top-of-the-line even in its day. The higher price range would give you something newer or a more fully equipped older rig. You can expect computer control, good filtering, solid-state, and an antenna tuner, but these features are not the latest and greatest.

Moderate Used Station

Used HF Transceiver	$300- $1,000
Used Antenna Tuner	$100
Multi-Band Wire Antenna	$75 - $150
Feed line	$50 - $100
Used Power Supply	$100
Misc	$100
TOTAL	$750 - $1,550

Specific Considerations
There have been hundreds of different radios produced in the last 15 years. I can't possibly list a fraction of the candidates.
- Check the reviews on eHam.net
- Look for features like a built-in antenna tuner and extra filters

ASSEMBLING AN HF STATION

- Ask the seller about included accessories (microphone, cables, original boxes, and manuals)
- Has the radio been in a smoking environment?

Moderate New Station

If you decide to go with all-new equipment, the price escalates. The following chart describes an all-new Ham station with moderate-level components. This isn't the most expensive gear available but is adequate, especially for beginners. After gaining some experience, upgrade, but many Hams stay in this category for years.

Moderate New Station:

Mid-range HF Transceiver	$500- $1,500
Antenna Tuner[28]	$100 - $250
Multi-Band Wire Antenna	$75 - $150
Feed line	$50 - $100
Power Supply	$125
Misc	$100
TOTAL	$950 - $2,225

There are options. Anything you can beg or borrow will reduce the initial cost. These figures are very rough estimates intended to get in the ballpark.

Specific considerations

The bottom of the price range will sport 100 watts, all-mode, HF only (160 meters to 10 meters), and no internal antenna tuner.

Moving up adds the antenna tuner and more advanced features like digital signal processing and filtering.

---

[28] New or used. Maybe, you won't need a separate tuner if your transceiver has an internal tuner. See more in the chapter on antennas.

ASSEMBLING AN HF STATION

Where to Buy

There are several highly reputable and very helpful dealers online. Visit their retail stores to spin the knobs on new gear. Most offer free shipping on orders over $100, and they will be glad to send a catalog.

Equipment Dealers Include
- Ham Radio Outlet at HamRadio.com Order online or at 13 stores across the country
- DX Engineering at DxEngineering.com online and in Tallmadge, OH
- Gigaparts at Gigaparts.com online and stores in Huntsville, AL, and Las Vegas, NV
- Ham City at HamCity.com online and in Gardena, CA
- R&L Electronics at RandL.com online and Hamilton, OH
- Quicksilver Radio Products at QSRadio.com
- Tower Electronics at PL-259.com
- Wireman at TheWireMan.com
- Digital radios at BridgecomSystems.com

USED VS NEW

Much like computers, radio equipment has undergone incredible technological advances. However, radio science hasn't changed, and many 50 to 60-year-old rigs are still in service. One can save money with used equipment, but what are the considerations when buying used?

Vacuum tube gear is a collector's item. Some of it still works just fine. I have a 60-year-old Drake transceiver that I enjoy very much. However, parts and tubes for old radios can be hard to find. Tube radios are projects like an antique car. I would not recommend you start with vacuum tube gear unless someone gives it to you.

ASSEMBLING AN HF STATION

Finding replacement parts for old gear can be a problem. By "old," I will arbitrarily say, 20 years. It's not that you can't find a particular resistor; it is the no-longer-made band switches, displays, and mechanical pieces that are "unobtainium." Older radios lose alignment, and parts may change value with time. Keeping up old equipment is a hobby unto itself, and unless you are comfortable troubleshooting and fabricating, you might better stay away.

Having warned off older gear,[29] I must add that many vintage transceivers still perform excellent service. Take one as a loaner or a gift but don't pay a lot, $250 to $350 maximum. Ensure it works because, even if parts are available, a professional repair (with shipping back and forth) might cost more than the radio.

Ten to fifteen-year-old gear is priced temptingly at a half, or less, of its cost when new. Future depreciation will be slow. It will sell for close to what you pay as long as nothing breaks. Check the reviews online (eHam.net). See if a particular issue plagues the model, such as failing displays, bad final transistors, or obsolete replacement parts. Older radios can be of real value if the radio works properly and doesn't stink from being in a smoker's shack.[30]

Five-year-old radios are practically new and sell for 75-80% of their original price.

You will get your best prices buying from a local Ham or at a hamfest. There was a time when you could trust a Ham not to sell a problem radio without full disclosure. I don't know if that courtesy exists any longer. Know your seller. Buying from a local Ham is probably safe. Don't expect a guarantee. If you want a warranty, buy something new.

---

[29] Sometimes called "boat anchors."
[30] The "shack" is a radio room.

ASSEMBLING AN HF STATION

Buying used radios online is riskier, though I have experienced good luck with sellers who had high positive feedback. PayPal and eBay have strong buyer protection policies, with recourse if the radio is not as represented, dead, or damaged on arrival. They can't help if the gear blows up after a week, even if you suspect the seller knew something was going wrong.

Other online sites don't have the same buyer protections, so understand the terms before paying. There are plenty of online scammers, and if the price is too good to be true, it is.

Try to borrow gear. The local club may loan equipment to new Hams. A member might have a radio sitting in the bottom of his closet that he would let you use. Finding equipment is another good reason to join your local club.

If you borrow, be a good steward and treat the gear carefully. Have someone show you how to use the radio and spend time in the instruction manual before doing any damage. The manual is often available online. Get a clear understanding of liability. Is the responsibility, "If it breaks in your possession, you bought it" or "I understand it is old and feeble, so whatever happens, happens?"

Agree, in advance, on the terms of the loan. I once let out a radio with no clear understanding, and it took me three years to get it back. If I loan again, I will take a picture of the borrower with the equipment and a sign that says, "Return by Jan 1, 2023." That gets the point across and helps everyone remember. On January 1, I can send him the picture as a hint. I am only half-kidding.

HF TRANSCEIVER CONTROLS

UHF/VHF radios have few controls, a channel selector, volume, and squelch. There might be a high/low power switch and a few buttons for quick access to memory channels.

The array of controls on a modern HF Ham transceiver is overwhelming. Large panels allow room for lots of knobs and buttons. Small transceivers have very few controls and rely on multiple button pushes or menus to reach the adjustable parameters. Here is a list of standard controls and what they do. I use the terms "switch" and "button." You flip a switch and push a button, but the function is the same.

AF Gain A volume control for the audio amplifier. Compare this to the RF Gain control for the radio-frequency stage amplification.

Antenna Selector A transceiver may come with more than one antenna jack on the back. Select the antenna from this switch or button.

AGC (Automatic Gain Control) Suppose you listen for a weak signal and have the gain (volume) turned up. Then a very loud station comes on. AGC protects your ears by leveling out the volume.

AGC has fast and slow settings. "Fast" recovers quickly, while "slow" will hold down the amplification longer. Use different settings depending on the conditions. Voice communications sound better with "slow."

Attenuation reduces all energy coming into the receiver, both noise and signals. It is handy for the lower frequency bands, where a high noise level can overwhelm the sensitive circuits. If the signal is above

HF TRANSCEIVER CONTROLS

the noise level, you will still hear it with attenuation dialed in.

Antenna Tuner Properly called a conjugate matching device, the antenna tuner adjusts the load to match the transceiver. There is a clattering of relays as an automatic tuner works to find the right solution. Don't worry; it is not broken.

Band Switch Rather than spin the frequency knob to go from 40 meters to 20 meters, one push changes the band, returning to the last frequency you used on the new band.

Compression Voice has loud and soft inflections as some words are naturally louder than others. Compression fills out the audio by boosting the less-strong words and syllables. Adjust the amount of compression to get that boost, but not so much as to add distortion.

CW Speed Most transceivers have a built-in keying circuit. This control adjusts the speed of the Morse Code sent from paddles.

CW Pitch The pitch is the musical tone. When set to 650Hz and the received audio signal is also at 650 Hz, you will be on the same radio frequency as the sender or "zero-beat." Pick a tone you like to hear and teach your ear to remember it.

Frequency Entry Enter the frequency on a keypad rather than spin a dial.

Filter Width Fixed or continuously variable filters restrict the range of signals heard, reducing interference.

HF TRANSCEIVER CONTROLS

Filter Shift (PBT, passband tuning) Alters the shape of the filter to favor frequencies above or below the center of the filter.

Mic Gain Adjusts the volume of the microphone input to the transmitter. Talk close to the microphone and adjust the mic gain per the transceiver's instructions. Most Hams "close talk" the microphone within an inch of their mouth. Speaking too far away and turning up the mic gain to compensate will also pick up background noise such as barking dogs or fans.

Message Playback Some transceivers have a built-in recorder for voice playback to record a message and play it back with one button push. A recorded message makes it easier to call and helps you get over mic fright. Pushing a button is less intimidating than calling.

Memories store frequencies and modes, making returning to a particular frequency easier. Set a memory slot for the band's top, middle, and bottom to move around without spinning the dial.

Mode Selects AM, USB, LSB, CW, and Digital modes.

Monitor Listen to the transmitted signal through this function. I keep the monitor volume low, so my ears stay sensitive to receive.

Noise Blanker Activates a circuit designed to mute the sound of repetitive noise such as a spark plug or electric fence.

Noise Reduction Filters out random noise. Very aggressive noise reduction can introduce distortion, so adjust by trial and error. There may be multiple settings to deal with differing conditions.

HF TRANSCEIVER CONTROLS

Notch Reduces or "notches" out an offending signal. Auto-notch will seek out and silence a carrier[31] near your frequency. Adjust a manual notch to do the same thing.

Power The on/off button, of course, but there is also a control to set how much power the radio transmits.

Pre-Amp An additional amplifier is applied to signals before they enter the receiver. Use this sparingly and only on higher frequencies with low background noise, or noise will overwhelm the receiver.

QSK When activated, your transceiver will switch from transmitting to receiving almost instantaneously. QSK is very helpful on CW as you can hear the receiver between sending your characters.

Receiver Incremental Tuning (RIT or Clarifier) Tunes the receiver without changing the transmit frequency. If someone calls slightly off frequency, use this to zero them in and "clarify" the signal.

Adjusting the VFO to make the incoming signal sound better would change your transmit frequency and sound weird on his end. Then he would adjust, and the two of you would chase each other all over the band. Transmit in one place and use the RIT to correct for the off-frequency signal.

Reverse To change from USB to LSB or to change the side to hear a CW signal. On CW, this is an effective interference fighting too. Reversing the side will make the interfering signal further away from the tuned frequency. On SSB, stick with the convention for the particular band. Listening to an upper

---

[31] If you are listening to SSB and someone decides to tune up near your frequency, you will hear the tone of his carrier. Notch will cut it out.

HF TRANSCEIVER CONTROLS

sideband transmission on a lower sideband setting does no good. It won't be understandable.

RF Gain Adjusts the gain in the radio-frequency section of the receiver. Turn the RF Gain down to hear the background noise just barely. This setting preserves the maximum dynamic range of the receiver. If a signal is below the noise, turning up the noise won't make the signal easier to copy. Turning down noise is also less fatiguing.

Spot Injects the audio tone selected with the CW pitch control to help zero-beat the other station.

Squelch Silences the receiver when there are no signals. UHF and VHF FM often use squelch. HF signals may be very close to the noise level and won't be loud enough to open the squelch. HF doesn't use squelch because it only triggers with signals substantially above the noise level.

Transmit Incremental Tuning Changes the transmit frequency while leaving the receive frequency the same.

Tuning step Determines how quickly the frequency changes as you spin the dial.

VFO Variable Frequency Oscillator is the frequency tuning control.

VOX Engages the voice-operated relay, avoiding a transmit/receive switch like the push-to-talk commonly found on a microphone. VOX facilitates hands-free operating.

VOX Delay Sets how long the radio stays in transmit after you stop talking. Setting the delay too short can be very distracting.

HF TRANSCEIVER CONTROLS

VOX Anti-Vox Noise from your receiver speaker can fool the VOX into thinking someone is talking and wants to transmit. The Anti-Vox control is negative feedback that cancels the received signal, so it doesn't trip the VOX into transmit mode.

This QSL card shows the front panel of an Elecraft K2 transceiver. It features multi-use buttons. Push activates one function. Push-and-hold activates a second.

HF TRANSCEIVER REAR PANEL

The typical UHF/VHF transceiver has little on the back panel, just the power connector, an antenna connector, a speaker jack, and perhaps a USB port to program or update the radio by connecting it to a computer.

The rear of a Ham HF transceiver can be as overwhelming as the front. If you're looking for a particular control and can't find it, check the back of the radio. There may be a few controls hidden on the back for functions that rarely change, such as microphone compression or VOX delay. Also, look for a trapdoor on top of the chassis.

The rear contains connections, discrete jacks, and cables for various functions, usually bundled in a DIN connector[32]. A DIN connector is small and round, with up to 14 pins (most are fewer). The plug has a key tab, so it will only fit the jack in one orientation. If there is a problem pushing in a DIN plug, the key isn't lined up. Back off, be gentle and don't force it, or you will bend the pins.

The rear of the equipment might be hard to access, and the only way to see it is to lean over the top. But then the labels are upside down. Take a picture or sketch for reference. Stick labels on the back panel upside-down, so you can read them while you hang over the top. The connectors all look alike otherwise.

---

[32] A DIN connector is an electrical connector that was originally standardized by the Deutsches Institut für Normung (DIN), the German national standards organization.

TRANSCEIVER REAR PANEL

Here's a list of what you might find on the back of an HF Ham transceiver:

Accessory Jack Control other equipment or pass audio. The various pins may allow additional audio inputs and outputs and packet or RTTY (Teletype) signals.

ALC Automatic Level Control interfaces with an amplifier ensure the transceiver doesn't apply too much power and overdrive the amp. If the amplifier senses it is over-driven, it changes the ALC voltage to reduce the transmitter's power.

Amplifier Keyer (may be called "key out"). A cable connected to an external amplifier to switch to transmit mode when the transceiver transmits.

Antennas Connections for the HF and VHF/UHF antennas.

Antenna tuner Interface with an external antenna tuner to put the radio in "tune" mode and send a low power signal to the tuner. A low power signal is easier on the relays and other components in the tuner. Once it has reached a match, apply full power.

Audio output Plug for an outboard speaker or headphones.

Fuse or circuit breaker Many transceivers have a safety fuse or breaker in addition to a fuse in the power cord. Keep a few spare fuses in your "go box."

Grounding lug To attach station grounds. See the chapter on "Grounds and Lightning Protection."

Key or paddles Input for CW operation. A key uses a mono plug with its two wires. Paddles use a stereo

TRANSCEIVER REAR PANEL

plug for ground, dits, and dahs. Most transceivers have separate jacks to accommodate both.

Line in and Line Out Fixed-level audio for digital modes and voice keyers or recorders. These may be a DIN plug or individual audio plugs.

PTT Separate push-to-talk switch. You may have a switch on your microphone that activates through the microphone jack. A separate PTT connector is for a foot switch or other hand switch.

RX Ant For a separate receiving antenna.

Serial/USB port so the radio and computer can communicate.

ANATOMY OF AN AMATEUR RADIO HF CONTACT

Always check the frequency before transmitting. Be sure to stay within limits for your license class. Extra Class licensees can operate in the lower portion of the band, and foreign stations on 40 meters converse where no US phone station can operate. Don't assume because you hear voices, you can transmit there as well.

Make sure the frequency is not already in use. Ask, "Is the frequency in use?" and identify yourself. Ask more than once.

What is the right way to start and sustain a conversation? You could start by calling "CQ." CQ is a general call to anyone, inviting them to join you. Never use CQ on a repeater. Just announce your call sign. "K4IA listening."

QSL card from Germany.

ANATOMY OF AN AMATEUR HF CONTACT

While new, I don't recommend calling CQ until you get comfortable conversing on the radio. Listen and find someone else calling CQ. Why?

Ham etiquette says the person who called CQ sets the tone and pace of the conversation. Admit it; you don't know what you are doing. Mr. CQ goes first, giving a signal report, location, and name. You give him yours. He might describe his equipment, and you reply about yours. Let Mr. CQ lead. No need to panic about, "What do I say next?" Follow the other guy's prompt and respond to him. Try to think of a question to keep the conversation going.

Keep your transmissions short. Regularly spoken conversation is one or two sentences, and then the other person speaks or interrupts. When transmitting, you can't hear the other side and could go on forever, as no one can interrupt. Long discourses are tough to follow. Don't be a hog. The other person should not have to take notes, so he remembers your 14 points before responding.

Exchanges are more interesting if they proceed like a regular conversation. The pattern should be brief two-ways and not extended monologs.

Answer a CQ by giving his call sign once and yours twice. Use phonetics with your call sign. "WA4TUF, this is Kilo 4 India Alpha, Kilo 4 India Alpha." You might add, "How copy?" "Over." or say nothing. Mr. CQ will hear the silence and know you have stopped calling.

A conversation does not require identifying on every "over." A simple, "What do you think, John?" is enough to let the other guy know it is his turn. Identify every ten minutes and at the end of your contact, as the Amateur rules require. The rule for GMRS is 15 minutes.

ANATOMY OF AN AMATEUR HF CONTACT

Once you get comfortable with HF conversations and are ready to lead one, call CQ. Before calling CQ, listen. Make sure the frequency is not in use. Sometimes, you can't hear both sides talking, so identify and ask, "This is K4IA. Is the frequency in use?" Listen for a few seconds and ask again.

Once sure you are on a clear frequency appropriate for your license, make a call. "CQ CQ CQ this is K4IA, Kilo 4 India Alpha, Kilo 4 India Alpha, calling CQ and listening."

I prefer several brief calls to one long one. Listeners will tire of a long CQ and spin the dial. If my short CQ doesn't raise an answer, I call again after waiting at least five seconds.

When there is an answer, acknowledge the station so he knows you are calling. "WA4TUF, K4IA returning. Thanks for the call. You're 58[33] in Fredericksburg, Virginia. Name here is Buck, Bravo Uniform Charlie Kilo. How copy?"

Notice, I use his call once. He knows his call sign and presumably will recognize it. I don't need to repeat my call. He knows it because he called me. I might repeat the signal report if conditions are poor but usually once is enough. I don't spell out Fredericksburg phonetically. It is long enough as it is. If the other guy doesn't get it the first time, he can ask. I spell out my name, maybe twice, if conditions are poor or the other guy sounds weak.

That initial exchange is the classic trinity of signal report, location, and name. Where it goes from there is up to you. On the next go-around, I might tell him where Fredericksburg is, my equipment, my weather, my other hobbies, what I do for a living, how long I

---

[33] A Signal Report. 1-5 for overall readability and 1-9 for signal strength.

ANATOMY OF AN AMATEUR HF CONTACT

have been a Ham or my favorite Ham Radio things to do. That would take several transmissions.

So far, I have only talked about myself. To be a good conversationalist, learn to listen. Ask the other guy about things that interest him. "My rig is a WizBang 4000 Pro. What's yours?" "I like to write books. What do you do when you're not on the radio?" Find common ground or learn something new.

Eventually, you will run out of things to say. How to end the conversation? You could take the coward's way out and claim the XYL[34] is calling, the dog is barking, or the phone rang. There is no need for deception. It is acceptable to say, "Thanks for the contact, John. I think I'll try to scare up a few more before I go to bed. Hope to hear you again soon. 73[35] and good night. WA4TUF this is K4IA clear on your final." Adding, "Clear on your final" lets listeners know you are not entirely done and discourages tail-ending interrupters. Ending a radio conversation is not nearly as awkward as walking away from a cocktail party windbag.

You will meet some very intriguing people. Included here are some interesting QSL cards to prove the point. I enjoy exchanging cards with Hams around the world and have shoeboxes full.

Take "Tex." Tex, W5 Big, Quick, Ugly was 101 when I talked to him on my commute home. He was quite a character and kept me in stitches.

---

[34] XYL is Hamspeak for wife.
[35] "73" is an old telegraph expression meaning "Best regards."

ANATOMY OF AN AMATEUR HF CONTACT

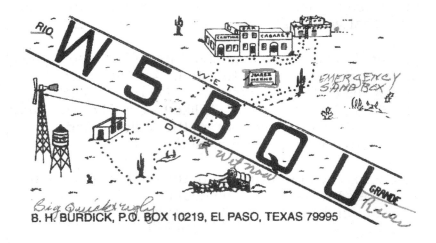

At the time we talked, Tex was reputed to be America's oldest active Ham. He is a Silent Key[36] now.

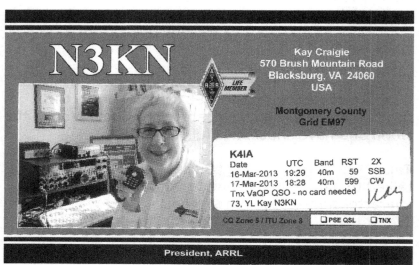

Kay was ARRL President when we met on the air.

---

[36] Silent Key refers to a deceased Ham. It is a throwback to the days of telegraph.

ANATOMY OF AN AMATEUR HF CONTACT

The first couple of letters identify the country. The number is an area within that country.

Who can resist this card from Northern Ireland?

ANATOMY OF AN AMATEUR HF CONTACT

A22 is the prefix for Botswana. A Japanese Ham was operating there. Hence, the call sign A22/JA4ATV.

ZS is South Africa. Art is licensed there and in Lesotho.

ANATOMY OF AN AMATEUR HF CONTACT

One of my favorite cards is from Kazakhstan.

Gyuri is QRL (busy) chasing QSL cards.

ANATOMY OF AN AMATEUR HF CONTACT

You can talk to the radio room of the *HMS Queen Mary, USS Wisconsin*, and many other vessels. The *USS Wisconsin* is berthed in Norfolk, Virginia.

"Big Mo" is stationed at Pearl Harbor, Hawaii. The Japanese surrender ending WWII took place on its decks.

AMATEUR RADIO RESOURCES

ARRL The American Radio Relay League has an old-fashioned name. It has been around since 1914. The League is a prolific publisher, advocate, and defender of Amateur Radio, billing itself as "the voice of Amateur Radio." The website, ARRL.org, is full of excellent articles and advice.

Magazines The two preeminent Ham Radio print magazines are *QST* and *CQ*. I enjoy them both. Both are available in paper or digital versions.
- *QST* is the official journal of the ARRL and comes as a benefit of membership in the American Radio Relay League) ARRL.org/membership ($49/year
- *CQ* is less formal and less technical but more fun. CQ-Amateur-Radio.com ($27/year)

Online
- eHam.net has articles, equipment reviews, classified ads, and news; the material is free, but they encourage contributions
- QRZ.com is also a free resource with news and an extensive call sign database; subscribers get to see more information
- Facebook has groups for every interest
- YouTube has videos of almost everything, including unboxings, reviews, and "how-to" instructions.

ASSEMBLING A VHF/UHF STATION

EQUIPMENT FOR A VHF/UHF STATION

Transceivers are the rule for VHF/UHF equipment. The transmitter and receiver are in one package. The transceiver can be an HT, base (desktop), or mobile (car-mounted) station.

The Handie Talkie or Walkie Talkie (HT)

HTs are the most basic VHF/UHF station, output 5 to 10 watts and get power from batteries. Over 5 watts needs more battery and will be heavier and bulkier. Many HTs are "over-hyped" and don't put out the stated power using their internal batteries, rated around 7 volts. Higher power requires a 14-volt source like a car's cigarette lighter or an external battery pack. The 7-volt version will deliver half the stated power.

Rating power in milliwatts is another example of advertising hype. A milliwatt is 1/1000 of a watt (.001 watts). Although the number looks large, a rating of 8000 mW is just 8 watts. I saw a handheld Chinese radio advertised as 8000 watts, conveniently omitting the "m" in "mW." This is an example of why you need to understand some electrical theory.

HTs use rechargeable batteries, but some manufacturers offer additional battery packs that accept standard AA batteries. Those are handy as a quick replacement for a rechargeable while charging.

Turn down the transmit power to the minimum needed to make contact to maximize battery life. The FCC rules say to use the minimum power necessary to maintain effective communications. This rule is to reduce interference and applies to all radio services.

ASSEMBLING A VHF/UHF STATION

Batteries last longer using an earbud instead of the speaker. The audio amplifier uses much more power when driving a speaker. Adjust the squelch such that you hear no noise without a signal. These suggestions reduce the radio's power consumption.

Plan for every HT to have three power sources: internal battery, backup battery, and wired. The wired connection can be a cigarette-lighter plug. Don't get caught short. Prepare for all three. A dead battery turns your HT into a brick.

The modulation mode used by HTs is FM. FRS and GMRS HTs operate on a single band. Some Ham HTs operate on a single band, 144 MHz, but most are dual-band, adding 440 MHz. The dual-band versions cost little more, so get a dual-band if your budget allows. Check your local conditions before you pay extra for a three-band Ham HT. That third band (usually 220 MHz) may not be active in your area. Lack of activity might be an asset if your group wants to remain discrete.[37]

GMRS and FRS HTs are channelized, choosing a channel, not a frequency. Turn to channel 3, for example. There is no need to program frequencies or repeater offsets.[38] The radio has that information stored.

Some imported radios advertise as "dual-band" and cover 136 to 174 MHz and 400 to 480 MHz, including Ham and other services. They would be illegal for GMRS use if not channelized.

Hams may not use their radios to talk to GMRS or FRS radios and vice versa. Legal radios do not transmit on

---

[37] Laws prohibit codes and ciphers. Use uncommon modes and frequencies to avoid eavesdroppers.
[38] See the later chapter on Repeaters

ASSEMBLING A VHF/UHF STATION

Ham and GMRS frequencies. Stay within your license class. Check the details before buying.

Ham radios are not channelized, and the frequency and tone combinations are almost infinite. Program Ham repeater frequencies, offsets, and tones through the HT keypad, but it is easier to use a computer program.

Find local repeater information with a smartphone APP called RepeaterBook or at RepeaterBook.com. Use the computer program that came with your radio, look for the free program, "Chirp," or invest in the excellent program and cables from RT Systems.com.

Accessories to complement the HT include:
- Programming cable to connect the radio to a computer
- Outside antenna[39]
- Full-size attachable antenna
- Extra rechargeable battery or AA battery carrier
- Earphone, because the internal speaker can be hard to hear in a noisy environment.

Base/Mobile Station

An HT alone is not enough radio. It may be handy for walking around, but you need something more potent for effective communications. That requires a Base/Mobile station. VHF/UHF gear is small and fits on a desk or mounts in a car. There is another chapter on mobile operation with more detail.

The typical Base/Mobile VHF/UHF rig will output 5 to 50 watts for GMRS and 50 to 100 watts for Ham. Some Ham models can operate as a cross-band repeater. Here is how that might work. The 440 MHz

---

[39] Your car is a metal box and if you try to operate an HT from inside the car, your signal will be very weak. Get an outside antenna. One type clips over your window.

ASSEMBLING A VHF/UHF STATION

band works better than 144 MHz for penetrating walls. When our local club supports the Special Olympics, we are inside a school building. An HT on 440 MHz communicates to a cross band repeater in the parking lot, running higher power and relaying to the big repeater operating on 144 MHz. This arrangement extends the reach of the in-school HT to cover the entire county. Operating cross-service with a GMRS radio is not legal and should not be possible if the radio past FCC approval.

CHOOSING THE EQUIPMENT

The big manufacturers are Icom, Kenwood, Yaesu, and Alinco. Motorola produces expensive commercial-grade gear.

There are a gaggle of inexpensive Chinese radios with names like Wouxun, Baofeng, TYT, Xiegu, Any Tone, and Hesenate. They look alike and probably come from the same factory. They have similar functionality, but the big four excel in spectral purity, ruggedness, and overall quality.

Radios are available through Amazon, but order from the retailers listed in the chapter on assembling an HF station for the best deals and service on Ham Radios. Avoid ordering any radio directly from China. The shipping and warranty hassles are not worth the few dollars you might save.

I will not recommend particular models or manufacturers. Manufacturers introduce new gear often, and it would be too hard to keep up. We all have our favorites, and some people like Fords while others favor Chevys. As my beloved Nana used to say, "That's why they don't make all ice-cream vanilla." Here are some criteria to apply when choosing equipment for a VHF/UHF Ham or GMRS station.

ASSEMBLING A VHF/UHF STATION

HOW MUCH DO YOU WANT TO SPEND?

UHF/VHF gear is less expensive than HF. The price is lower because it handles less power and uses limited frequencies and modes. That said, it can be expensive.

New digital modes on Ham radio add considerable cost and functionality to a basic radio. They go by names like D-Star, DMR, and System Fusion. No one system is supreme. Remember the battle between Sony Betamax and VHS tapes? Investigate before committing. Check RepeaterBook.com to see what your area supports. It makes no sense to spend big dollars on a digital radio if there is no digital activity or you have the wrong format unless you are trying to be discrete and avoid everyone else.

Start with a simple single-band FM radio. Most Amateur Radio activity is on two-meter FM. Check with the locals before spending money on bands and features you will never use.

Used Equipment
Basic used VHF/UHF Ham equipment is not expensive. By "basic," I mean one or two bands, FM, either an HT or a small mobile rig. A club member might give or loan something. Used CB equipment may have been abused, while GMRS equipment may have spent its life in the original box.

Things to watch when buying used:
- HT battery is dead or has little life left; a replacement might cost as much as paid for the radio if one is even available
- Missing manual
- Broken antenna
- Missing parts such as the microphone, mobile mounting bracket, charger, or power cable.

ASSEMBLING A VHF/UHF STATION

Amazon and eBay have replacement parts. Find manuals on the manufacturer's website or the Internet.

New Equipment
Everyone needs at least one HT. Cheap Chinese versions start at less than $50. Get a cheap throwaway HT to start. You won't cry if it is lost or damaged.

A 5-watt dedicated desktop or mobile GMRS radios transceiver is around $100, 15 watts is $150, and 50 watts is $250 currently on Amazon. These include a magnetic mount antenna to stick on the car. I wouldn't bother with the 5-watt version. Use an HT.

A single-band Ham desktop or mobile VHF 65-watt radio is around $150. Dual-band base/mobile Ham radio (not digital) is around $250. Add $50 for an antenna to stick on the car.

To use any of these radios at home, you will need an outside antenna ($50 to $100), cable to feed it ($75), and a power supply ($100). These are very rough estimates, as every installation is different.

RECOMMENDATIONS
- At least one HT for GMRS and one for two-meter Ham. FCC regulations do not allow Ham and GMRS in the same radio
- A 15 to 50-watt GMRS or two-meter Ham base station will provide a more robust link
- Use the radios with a high antenna
- Use a directional antenna to extend the range
- Buy new. You might save a few dollars on a used unit, but you will not be happy if it doesn't work when the need arises.

REPEATERS

A VHF/UHF Amateur or GMRS radio can talk directly line-of-sight. VHF and UHF radio waves do not bounce off the ionosphere. They will bounce off buildings, and if you're high enough or conditions are just right for tropospheric ducting, you might communicate further. However, tropospheric ducting is rare. We need some help with reliable communication, and a repeater is an answer.

A repeater simultaneously receives and retransmits a signal. The repeater "repeats" the signal to relay over a longer distance.

Repeaters need power to operate, and the premise of this book is "no power." I include the discussion of repeaters because many have backup power supplies. Our club's Ham repeater is co-located with the county emergency services and assured of power as long as they have it. That won't help if the antenna blows down. Use repeaters but have another plan when they fail or get busy in a disaster.[40] A repeater tied to emergency services will not be available for your personal use.

A local club builds and maintains the repeater with member dues, which is an excellent reason to support your local club.

Duplex mode Simplex communication describes a station transmitting and receiving, but not at the same time. Repeaters operate in "duplex" mode, receiving and transmitting simultaneously. The repeater listens and talks on two different frequencies to prevent self-interference. It cannot transmit and receive on the

---

[40] Three is two, two is one and one is none. This adage applies to all preparedness planning.

REPEATERS

same frequency, or the repeater would interfere with itself.

Offset "Repeater offset" is the difference between a repeater's transmit and receive frequencies. A standard repeater offset in the two-meter Amateur band (144 MHz) is plus or minus 600 kHz. For example, my local repeater receives at 147.615 MHz and retransmits on 147.015 MHz. To access the repeater, you would tune your receiver to 147.015, so you can hear the repeater and set the offset on your radio to transmit +600 kHz so the repeater can hear you. Store this information in a memory channel.

On the 440 MHz Amateur band and GMRS, the repeater split 5 is MHz from the receive frequency. GMRS radios do this automatically when tuned to channels 15 through 22, reserved for repeaters.

Range Since VHF/UHF communication is line-of-sight, the higher you are, the further you can see, and the further the radio waves will reach. That is called the radio horizon. A repeater antenna will be on a very high site atop a tall building, water tower, or mountain.

Besides the advantage of a high antenna, the repeater transmits more power than the typical handheld or car-mounted radio. The combination of extra height and power stretches the repeater's range out to perhaps 50 miles. That radius would include almost 8,000 square miles. Your useful range will depend on your HT or mobile radio's ability to reach the repeater.

Linked network A linked repeater network is a network of repeaters where the others repeat signals received by one. Linking expands coverage.

Memory channels A way to enable quick access to a favorite frequency on a Ham transceiver is to store the

REPEATERS

frequency in a memory channel. The radio will have memory channels, so once you set the frequency and offset data in the memory, you do not have to re-enter it every time you turn on your radio or switch to a different repeater. GMRS radios only operate on set channels with the memory function built-in.

Tone codes Many repeaters use a sub-audible tone as a "key" for access to prevent interference with other repeaters. "Sub-audible" means too low in pitch for you to hear. Program the sub-audible tone in a memory channel with the other repeater data.

Continuous Tone Coded Squelch System (CTCSS) is the term used to describe the sub-audible tone transmitted with the voice audio to open a receiver's squelch.

If you're trying to access a repeater and failing, chances are you may not have the proper CTCSS, DTMF, or DCS tone for access. DTMF is the tone generated by the keypad. DCS[41] is digital. Don't let all those fancy names fool you. They are just another way of saying you may need a particular tone to unlock the repeater.

A repeater will identify in Morse code every ten minutes. If you can hear a repeater's output but can't access it, a reason might be:
- Improper transceiver offset
- The repeater may require a CTCSS tone
- The repeater may require a DCS tone.

APPS for iPhone and Android use the GPS on your phone to locate nearby repeaters and provide access tone information. Look for an APP called "RepeaterBook." Write these down or program the

---

[41] Digital code squelch.

REPEATERS

radio. You may not have access to the Internet in a disaster.

Repeaters can also provide an on-ramp to the Internet, and the term "gateway" describes a station used to connect to the Internet.

The Internet Radio Linking Project (IRLP) connects Amateur Radio repeater systems via the Internet. Repeaters linked by IRLP use DTMF tones generated by the transceiver keypad, just like a telephone's tones. DTMF stands for Dual Tone Multi-Frequency. Dial in the "phone number" of a distant repeater, and your signal will pop out of that repeater, even on the other side of the world. Dial 5600, and the signal will travel over the Internet to a repeater in London.

Echolink is another Internet linking protocol.

Repeaters are a backup, not a primary mode for emergency services. Our scenario assumes no power. Even if there is power, wind or flying debris can damage the repeater antenna. The repeater may suffer damage from lightning, EMP, or other misfortune. Competing users may overwhelm the repeater. Don't put all your eggs in one basket.

RECOMMENDATIONS

- Repeater access is tricky and might require some trial-and-error to get right; don't wait until it is panic time
- Repeaters fail, so prepare and practice to also operate in simplex mode
- Preparation requires planning, practice, and coordination. There is more in the chapter on Operating Tips and Strategies.

ANATOMY OF A REPEATER CONTACT

First, check that the frequency is not in use. Listen before you leap. Silence does not mean the repeater is not in use. There may be many stations on standby during an emergency.

Do not call CQ on a repeater. That's only for HF. To invite a general call, say your call sign. "K4IA listening." To start contact with a specific station, call him. "WA4TUF, this is K4IA. You out there?" GMRS operators identify with their call sign.

Don't make unanswered calls more than twice without waiting a few minutes. If there is anyone out there who wants to talk, they will answer. Repeated calling is begging and annoying.

To join an existing conversation, wait for an appropriate pause in the conversation and say your call sign. You'll be acknowledged and invited to join. Consider the conversation before calling. If the participants are in a deep discussion, don't interrupt to comment on something off-topic. That's rude.

"Break Break" is for emergencies only. Don't use it to join a conversation.

Repeater transmissions are short and conversational, not monologs. The repeater will time out and drop your transmission if you go on too long (usually defined as more than a minute). The time-out is to allow another caller to join. Pause between transmissions to listen for anyone who might want to join the conversation or report an emergency.

Identify with your call sign every ten minutes when transmitting and at the end of your conversation. The interval is fifteen minutes for GMRS transmissions.

MESH NETWORKING

Amateur mesh networking is an application that creates a network over Amateur Radio. The nodes interconnect in a spider web fashion and route the message along whatever path it might need to get to the final destination. If one node drops out or gets overwhelmed, the software sends the message through other nodes to reach home. The Internet is a wired mesh network. Amateur operators are creating a wireless Internet. You only need the entry-level Technician Class license to take part.

Wireless computer routers share frequencies with a portion of the 2.4, 3.4, and 5.8 Gigahertz (GHz) Ham bands. Some industrious Hams, exercising what I call "hamgenuity," modified router firmware to adapt it to mesh networking. They also boosted the router output. Hams can attach amplifiers and gain antennas, significantly increasing the range of an individual node.

Mesh networking is relatively inexpensive ($150 per node) and can provide a robust data network. The router draws minimal power and can run on battery power charged by a small solar panel. A combination of fixed and portable stations could cover a wide local area, with additional nodes connected to the Internet.

Anything "data" can travel over a mesh network, including email, video, or voice-over-Internet (Skype) audio.

The Amateur Radio Emergency Data Network (AREDN) is one group spearheading the effort. They define their responsibility as, "The AREDN™ development team strives to create quality software releases for use on commercial-off-the-shelf (COTS) devices with a primary focus on meeting the needs of emergency

MESH NETWORKING

communications data networks." The website is AREDN.org.

There has been recent controversy over using the group's name and acronym, which may hinder further development.

Another group promoting mesh networking has a website, Broadband-Hamnet.org. Their product is "a high speed, self-discovering, self-configuring, fault-tolerant, wireless computer network that can run for days from a fully charged car battery, or indefinitely with the addition of a modest solar array or other supplemental power source. The focus is on emergency communications."

It helps to have a computer networking background when setting up a mesh network, but further development promises to make the system plug-and-play.

Mesh networks show great promise for groups wishing to set up a wide-area wireless network designed to operate without mains power. Mesh networks would be ideal for offering broad coverage in a disaster area when all else fails.

OPERATING TIPS AND STRATEGIES

A communications plan should embrace multiple approaches. CB, GMRS, and Ham VHF/UHF are the obvious choices for local contacts, and longer distances require HF and a Ham license.

COMMUNICATION PLANNING

There are millions of permutations of time and frequency. Before HMS Titanic sank, there was no accepted radio protocol. The US and Europe heard Titanic's distress signals, but the nearest ship's operator had turned off his equipment and gone to bed. International reforms established designated listening times when all transmitting stops, and ships listen for distress calls.

Preppers need a similar system, and one is the 3-3-3 rule.

3-3-3 Rule The 3-3-3 rule is simple:
- Turn on your radio every 3 hours
- Listen for at least 3 minutes
- On channel 3 or the designated Ham frequencies
- Follow the same schedule if you need to call.

The start time is noon, so listen at noon, 3 PM. 6 PM, etc. Listen for at least three minutes on channel 3 of FRS, MURS, or CB or the designated Ham frequencies. Appendix D lists channels and frequencies.

Arrange a meeting schedule (time and frequency) for your inner circle. You could choose a different protocol, but the critical point is to have a plan so everyone can reasonably expect to be on the same frequency at the same time.

OPERATING TIPS AND STRATEGIES

The advantages of the 3-3-3 rule are:
- Easy to remember
- Short operating periods conserve battery life
- Everyone is on the air at the same time and frequency
- Sets a window of 8 times a day to call
- You don't have to call for hours, hoping someone will tune in to hear you
- The 3-hour schedule allows time to change locations for better reception.

The 3-3-3 rule requires only 24 minutes of listening each day. If the battery is good for ten hours, it will last up to a month.

NETS

A net is a group of users. They can operate on HF or through repeaters. Informal nets are roundtables, where the microphone passes around: Joe to Moe to Larry. Everyone gets a turn and passes to the next participant.

A formal net is "directed." A net control station (NCS) manages the session. He would first call the net to order, reciting a monolog of the net's purpose and protocols. Then, the NCS will ask for check-ins. First, come stations with emergency traffic (messages), then mobile or low power stations, then general check-ins. The NCS will recognize the individual stations and ask for their comments, one at a time. If two stations need to converse, NCS will send them off the net frequency, so it remains clear for net operations.

If you hear a net operating, check-in at the appropriate time but standby until called upon.

OPERATING TIPS AND STRATEGIES

PHONETICS

Standard phonetics cut through noise and interference and are easier for others to understand. Notice, I said, "standard phonetics." Cutesy or non-standard variations are confusing.

Here is the word list adopted by the International Telecommunication Union:

A--Alpha	**J**--Juliet	**S**--Sierra
B--Bravo	**K**--Kilo	**T**--Tango
C--Charlie	**L**--Lima	**U**--Uniform
D--Delta	**M**--Mike	**V**--Victor
E--Echo	**N**--November	**W**--Whiskey
F--Foxtrot	**O**--Oscar	**X**--X-ray
G--Golf	**P**--Papa	**Y**--Yankee
H--Hotel	**Q**--Quebec	**Z**—Zulu
I--India	**R**—Romeo	

The FCC initially assigned me the call sign KG4CVN. "G," "C" and "V" sound alike, especially in noise or fading conditions. So do "N" and "M." Golf, Charlie, and Victor don't sound alike. No one can confuse November and Mike.

HOW TO USE A MICROPHONE

You may think this is elementary, but you will hear plenty of poor mic skills. One example is holding the mic too far away. The result is the mic picks up the background noise. Another problem is insufficient voice volume. Talking too quietly doesn't "fill" the signal and is hard to understand. Yelling in the mic causes distortion and strains your voice.

When calling, be clear and crisp. Don't mumble, don't shout, and don't stretch out the phonetics. Keeeeellllooo, for K, is not easier to understand than Kilo.

OPERATING TIPS AND STRATEGIES

Slow down and enunciate clearly. The message will get through quicker if you don't have to repeat yourself three times. It annoys me when someone leaves a lengthy message on my voice mail and speeds through their phone number at the end, "5407858122." Avoid confusing fillers such as "ya know," "err," and "uh."

Get close to the microphone and speak in a normal tone of voice. Practice and ask others how you sound. Adjust the microphone's gain and compression controls if there are any.

ENCRYPTION AND CIPHERS

Communication laws prohibit the use of ciphers or codes to obscure a message. You can speak a foreign language but must identify in English. Morse code is not a cipher because it is widely used and understood. The developer of a digital mode is required to make the decoding key public.

A prohibited cipher is seemingly random numbers or letters that only make sense after conversion. The listener shouldn't need a Little Orphan Annie decoder ring.

Trigger words or phrases would seem to slip by the rules. For instance, "The cow has left the barn." is not a cipher but could mean, "I am on my way." Ham rules say you cannot obscure the meaning of a message. The GMRS rules prohibit messages with hidden meanings and ban such subterfuge.

OPERATING TIPS AND STRATEGIES

Don't be an elephant, stomping over everyone attempting to make contact.

Don't monkey around while preparing.

ELECTROMAGNETIC PULSE (EMP)

An EMP or electromagnetic pulse is scary. Electromagnetic energy induces a current in wires and components. If the current is strong enough, it can damage electronic equipment. The sun, lightning, or a nuclear explosion could generate such a pulse. The fear is that an EMP will destroy electronics, satellites, GPS navigation systems, and undersea cables. In 1989, Quebec suffered an EMP-induced electrical system failure leading to a nine-hour blackout.

The most famous electromagnetic pulse of the modern era, the Carrington Event, occurred in 1859 when a huge solar storm sent a massive electronic energy burst to Earth. The northern lights, usually limited to areas around the poles, were visible almost to the equator. Telegraph operators reported sparks across their keys and electric shocks. The Carrington event was enormous, but tree-ring evidence suggests more significant events have occurred in the past.

Local sources can generate an EMP. I experienced EMP damage when a nearby lightning strike destroyed my receiver, even though the lightning did not hit my antenna. Even wind blowing across an antenna can generate enough static electricity to cause harm. You can't do much to protect the power grid, but you can protect your equipment.

Disconnect The examples' common denominator was a long piece of wire: an antenna, telegraph, or power line. The energy from the pulse transfers to the wire. Step number one to avoid EMP damage is to disconnect equipment completely when not in use. A pulse can't travel down the antenna wire into a radio if you disconnect the antenna. It can travel through the house mains to damage a radio only if the radio is plugged in. Disconnecting is a good habit to cultivate.

ELECTROMAGNETIC PULSE

Shield Another defense against EMP is to shield equipment by enclosing it in metal. The radio's metal case is a shield, and desktop radios have a ground lug (connection). If there is no ground lug, use an existing screw on the chassis. Scrape away any paint to ensure a good connection. Then, connect the case to ground, as described in the next chapter on grounding. If a ground wire is not practical, say with an HT, store the equipment in a metal trashcan and ground the can.

Predicting an EMP generated by an enemy attack is difficult, but we get notice of solar and lightning related EMPS. When there is a solar event, the first manifestation is X-rays from the solar flare. X-rays travel to Earth in 8½ minutes, the speed of light. They disrupt the ionosphere and shut down long-distance radio communications. The effect only lasts a day or two. Scientists can predict what is to come based on the intensity of the X-Rays.

A coronal mass ejection accompanies the flare. The CME is a far more damaging burst of plasma, hot magnetically charged gas, and particles that take several days to reach Earth. If Earth is in the material's path, we get auroras. If severe enough, it can also damage satellites and disable electrical grids.

You can monitor solar activity at SpaceWeather.com (assuming you have Internet).

NOAA weather broadcasts or the sound of thunder warn us of lightning. If you can hear thunder, you are within the possible lightning zone — time to shut down and disconnect. Don't jeopardize your life and equipment for a few extra minutes of operating time.

GROUNDS, LIGHTNING AND EMP PROTECTION

A typical station has three "grounds." An electrical safety ground, a lightning ground, and a radio frequency (RF) ground. They fulfill three different functions but operate in harmony. Modern electrical codes and good engineering practice require all grounds to be interconnected. The fundamental concept is to avoid any difference in electrical potential between components, which means minimal resistance to both AC and DC.

ELECTRICAL SAFETY GROUND

The electrical safety ground is the third hole in an electrical outlet and connects the green wire to the electrical panel ground. The plug wire connects to the equipment's case, whether a radio or a dishwasher.

We call this a "safety" ground because the purpose is to trip the circuit breaker if voltage appears on the equipment's chassis. The safety ground protects against electrocution. Do not defeat the safety ground protection with an adapter plug or any other gimmick!

LIGHTNING AND EMP GROUND

Thor's hammer carries an incredible wallop. A thundercloud can hold over 100 million volts of potential power and typically generates 5,000 to 20,000 amperes of current. A direct hit will be devastating. The lightning ground deflects power from a lightning strike by dissipating energy on an antenna or power line before it enters the house.

To protect against EMP, prevent it from entering by disconnecting. Additional protection comes from grounding and bonding equipment to provide an

GROUNDS LIGHTNING AND EMP PROTECTION

alternate route for the pulse that goes around the equipment rather than through it.

Lightning also creates an electromagnetic pulse. An indirect or nearby hit can induce damaging amounts of current into antennas and equipment.

For protection, reduce the resistance, and therefore the voltage difference between, around, and through the equipment and the home's electrical system. The voltage potential on the entire system should rise and fall together like a boat riding over a wave. Enormous amounts of current will flow if there is even the slightest difference in resistance between a station ground and the lightning ground.

For robust solutions that commercial installations use to survive direct strikes, search for, and download the Polyphaser "Lightning Protection and Grounding Solutions" PDF book.

Give lightning a path to ground outside the home. A lightning arrestor in the feed line and attached to a ground rod is just a start. Multiple 8-foot ground rods, connected in parallel, are better than one.

Lightning arrestors provide limited protection against nearby strikes but may bleed off potential charges to prevent a strike. The arrestor has a gap that will allow energy on the feed wire to jump to ground. A direct hit will destroy the arrestor. A few nearby strikes can damage the gap. Inspect and replace lightning arrestors often.

Putting an antenna switch in the "ground" position is not enough. The antenna may be "grounded," but all the other cables coming into that switch are connected, and the current will flow back to and through the connected equipment.

GROUNDS LIGHTNING AND EMP PROTECTION

Disconnect an outdoor antenna when not in use. I mean entirely disconnect, unscrew the cable, and don't leave it lying on any equipment or near a ground path. That lightning bolt travels thousands of feet and can readily jump a few more to get to your equipment or an electrical outlet. Don't rely on a lightning arrestor and don't rely on a switch for protection.

Summer thunderstorms are not the only source of damaging energy. Static electricity due to wind or rain can also cause damage. Dry winter air encourages static charges on the antenna. The components in our receivers handle millionths of a volt, not millions of volts. Disconnecting antennas is a good practice during all seasons.

Unplug the power supply and other equipment to protect against a power surge on the AC line as an additional precaution. I lost a very expensive amplifier to a power surge. Even though the amplifier was "off," the power supply was "on." When the lights flickered, the power supply fried. It was not repairable. Many electronics, including televisions, have energized circuits that run even though the device is off.

RF GROUND

We often think of "ground" in terms of direct current or alternating current at house frequencies, 60 Hz. Connecting a wire to a cold water pipe may be a safety ground,[42] and it may be of limited help to dissipate lightning surges, but it is not an adequate RF (radio frequency) ground.

"Ground" is not some mysterious pit to dump energy. At radio frequencies, every piece of wire has

---

[42] No longer recommended, as the water line outside the house could be non-conducting plastic pipe.

GROUNDS LIGHTNING AND EMP PROTECTION

inductance and capacitance, called "impedance." Impedance is resistance to alternating current. A ground wire may not be at RF ground potential. A good RF ground is a low impedance path for RF energy to flow around, not through equipment. If you experience "mic bite" or a burning sensation when touching a chassis while transmitting, that is RF. You might also notice distortion in the audio signal. The computer might reboot for no reason, or the radio display could go haywire.

Running low power, as in GMRS service, might not produce noticeable "mic bite" or burning sensations, but you could still have excess RF floating around on the equipment, distorting the signal.

There are many ways RF can get in radio equipment. One is an unbalanced antenna system. Even a balanced antenna such as a dipole can become unbalanced because of objects within its field, such as uneven terrain or metal siding. Another cause is a feed line that is not perpendicular to the antenna. Balanced feed lines cancel RF, and unbalanced feed lines radiate.

The antenna may induce RF in unwanted places. House mains, telephones, alarms, televisions, and TV cables all have wires that act as antennas. Various components use wire, such as the computer keyboard, mouse, video display, microphone, coax, and speakers. These wires are antennas. If they pick up the signal from a transmitting antenna, they will add noise and bring RF wherever they connect. Currents on the wires will also radiate, causing RFI (Radio Frequency Interference) in nearby equipment.

Keep the portable antenna away from the base station. I would want at least twenty feet separation at 50 watts. Don't mag-mount an antenna on the transceiver case.

GROUNDS LIGHTNING AND EMP PROTECTION

Other than distancing, there are three ways to control RF on wires and feed lines. Sometimes it takes a combination of distancing and all three.

Chokes A choke is a coil that increases the impedance (resistance to RF) to choke off RF. Enhance the coil's effectiveness with a bar or donut (toroid) of clay and metal. There are different metal mixes chosen for the frequency being choked. Mix 31 is used below 10 MHz. Above 10 MHz, mix 43 prevails. VHF/UHF uses mix 61.

Wrap the wire through or around the choke as many times as will fit. The impedance (resistance to RF) goes up with the square of the number of turns through a toroidal choke. Go through one choke two times to get four times the impedance. Go through two chokes once and only get twice the impedance. More turns are better than more chokes.

Here is an excellent article by K9YC, Jim, dealing with RFI and chokes.[43]

Twisted pair The second solution to RF induced in wires is to use twisted-pair wire. The twist will cancel any induced currents. Choose a twisted-pair cable for interconnections and stereo speakers.

Bonding The third solution is bonding. Bond each case to the others to provide a low impedance path for RF around the equipment rather than through the equipment. If RF has two paths, one through the equipment and one low impedance around the

---

[43] www.audiosystemsgroup.com/RFI-Ham.pdf

GROUNDS LIGHTNING AND EMP PROTECTION

equipment, most energy will travel on the low impedance path around, not through.

Connect all chassis: radio, antenna tuner, computer, and amplifier with a ground strap to provide a low impedance path from chassis to chassis. That encourages RF to stay off the interconnecting cables and out of the electronics. Also, run a strap to the shack's common ground connection going to the grounding rod. Use multiple connections between the equipment and station grounds. Use a bonding strap parallel to any cables or coax running between components.

RF travels on the surface of a conductor because of a phenomenon known as "skin effect." Grounding strap is flat, not round like wire, so it has more surface area. Therefore, the strap is a better RF conductor. The skin effect means the strap can be thin. A solid flat strap is a marginally better conductor than braided, but a braided strap is more flexible. Avoid braided for outdoor use because the braid corrodes easily.

To attach the strap to a chassis ground lug, poke a hole through the braid or drill a hole in the solid. Sandwich the strap between a couple of large washers and secure it with the nut on the equipment's ground lug. To splice straps, use the same technique, and sandwich them between two large washers.

Finally, electrical codes require bonding the RF, safety, and lightning grounds together. There should be no difference in potential between the various parts of the grounding system. Any difference in potential will cause current flow between the components, bringing hum, buzz, RFI, or damaging current from EMPs.

GROUNDS LIGHTNING AND EMP PROTECTION

The only thing more dangerous than improper grounding is crocodile taunting, as practiced here in Chad, North Africa.

Skydiving would be a close second to crocodile taunting.

ANTENNAS

There is no more contentious area of communications debate than the choice of antennas. Ask five Hams a question to get seven different opinions and maybe a fistfight.

A better antenna is the single most important investment you can make to improve your signal. Adding an external antenna will enhance CB, GMRS, and Ham signals. FRS radio can't use external antennas. A better antenna helps both transmit and receive, so money spent on the antenna system is very beneficial.

There are thousands of antenna designs and hundreds of books and articles on the topic. I am showing The Easy Way to understand some antenna theory.

Don't suffer paralysis by analysis. The antenna you have will contact lots of stations. The perfect antenna you are seeking but still haven't installed won't contact anyone.

THE DIPOLE

We begin with the dipole. Understanding the dipole is the key to understanding other antennas.

A dipole is a half-wavelength long, cut, and fed in the middle. Each side is a one-quarter wavelength long at the design frequency.

You can roll your own dipole easily and economically. Measure and cut two wires according to the formula 234/frequency = length in feet. Appendix A is a chart for each half of a dipole. That distance is the length of wire from the center insulator to the end insulator. Cut long and trim for results because removing excess wire is a lot easier than soldering on a longer piece.

ANTENNAS

As you can see from the chart, HF dipoles can be very long. VHF and UHF antennas are much shorter and easier to build.

Attach one end of each quarter-wave wire to each side of the center insulator. Feed in the middle by soldering the coax[44] center conductor to one side of the antenna and the coax shield to the other side. Support the coax feed line on the insulator with a strain relief fitting or wrap the coax around the center insulator. Soldered connections aren't strong enough to hold the weight of the coax. Pre-made center insulators with a standard SO-259 coax jack save the bother of soldering and provide strain relief.

Put another insulator on each end of the wire and attach the hanging rope to the end insulators. UV-resistant antenna rope, sometimes called "Para-Cord," is best. Sunlight will destroy other materials in a few months. I have had antenna rope up for over ten years without a problem. My book "How to Get on HF The Easy Way" will show you the tricks.

Trim the antenna for minimum SWR[45] using an antenna analyzer or SWR meter. Trimming is unnecessary at VHF/UHF frequencies as the changes are minimal.

Don't obsess as long as the SWR is below 1.5:1. Even 2:1 and higher is acceptable if your radio does not fold back power. If the SWR goes up with a higher frequency, the antenna is too long; wrap a bit of the wire back onto itself. If the SWR goes down at a higher frequency, the antenna is too short.

---

[44] Coaxial cable. Feed line is in the next chapter.
[45] Standing Wave Ratio is a measure of how well the system is matched. A reading of 1:1 is ideal.

ANTENNAS

Appendix A is a chart for various frequencies. If your antenna's lowest SWR is at 13.9 MHz (202 inches) and you want to move it up to 14.2 MHz (198 inches), the chart says to subtract the difference, 4 inches, from each leg of the dipole. The easy way to do that is to pull 4 inches of wire through the insulator and wrap it around itself at the end of the antenna. Shortening an antenna is easy. It is harder to add wire, so start a little long and wrap the excess back.

The antenna will move up in frequency when raised higher, so don't check the SWR on the ground and expect it to stay the same in the air.

Assuming choice, which direction should the antenna hang? Horizontal wire dipole antennas radiate off their sides, which is where they have gain (enhanced signal). If the antenna points north/south, it will radiate the best east/west. The pattern for horizontal antennas hung lower than a half-wavelength high becomes omnidirectional, favoring no particular direction.

VERTICALS

Not everyone has trees or other places to hang wire dipoles. Never use a utility power pole as it carries lethal voltages. Perhaps a vertical antenna is an answer.

The quarter-wave vertical is a cousin of the dipole. It is half a dipole, and something else provides the other half. The other half can be the ground, wire radials, your body holding an HT or the car chassis in a mobile installation.

A vertical can be a good antenna, and it is also relatively stealthy if you are concerned about neighbors and restrictive covenants. It can also use a tilt-over ground mount. Tip the antenna over when

ANTENNAS

not in use — that way, the petunia police from the homeowners' association won't notice.

On VHF/UHF, the antennas are short, and there shouldn't be much trouble concealing one. Get it high. Verticals are also easy to car-mount. If you don't want to drill a hole in the car, use a magnetic base or trunk-lip mount. Mounting the antenna in the center of a vehicle roof provides the most uniform radiation pattern. Mounting an antenna on the corner of a car changes the pattern, favoring specific directions. Since the car is moving, the results are unpredictable. If there is no choice, do what you can. It will work just fine.

A variation on a mobile antenna for VHF/UHF is 5/8 wave long instead of 1/4 wavelength. The extra length changes the antenna pattern to provide a lower take-off angle, concentrating the signal along the horizon. Therefore, another mobile station is more likely to receive the signal.

Another variation used for HF is an end-fed wire (EFW). It only requires one support fed at the bottom with a matching transformer. This design is popular with portable operators.

POLARIZATION

The orientation of the electrical field describes a radio wave's polarization. Radiation is the strongest broadside of an antenna. Whichever way the antenna is mounted or facing, most energy will be broadside to the wire and not off the ends.

The wire is horizontal for a simple dipole mounted parallel to the Earth's surface; the radiation is the strongest broadside, and the radiation is horizontally polarized.

ANTENNAS

A vertical antenna has an electrical field perpendicular to the Earth. Vertical antennas also radiate broadside, and since the antenna is vertical, they are vertically polarized. Mobile FM uses vertical polarization because that is how a car antenna fits and how you hold an HT antenna, up-and-down. Repeater antennas are vertical to match the mobiles.

Signals are significantly weaker when using different polarizations. A horizontal antenna talking to a mobile vertical or a vertically polarized repeater could be 100 times weaker. Don't hold your rubber ducky sideways, as that will make the antenna horizontally polarized.

When a signal bounces off something, it becomes wobbly like a tumbling hula-hoop. It becomes elliptically polarized (random both ways), and it doesn't matter how the antenna is set up. Skip signals are elliptically polarized, and you can use either a vertically or horizontally polarized antenna.

NVIS ANTENNAS

NVIS stands for Near Vertical Incidence Skywave. A low HF horizontal antenna[46] will direct the energy straight up or at a very high angle. When the HF signal hits the ionosphere, it reflects almost straight down. VHF and UHF signals don't reflect, so NVIS is suitable only for HF.

NVIS antennas provide good short-range (local to 300 miles) HF coverage. HF using an NVIS antenna will reach out further than a UHF/VHF radio. NVIS on HF is the best choice to contact the state capital 100 miles away.

---

[46] How low? Less than half as high as one side of your dipole is long.

ANTENNAS

ANTENNA GAIN

A directional or "gain" antenna concentrates the signal in one or more directions. The concentration of power can significantly enhance the transmitted signal and reception.

When analyzing gain figures posted for an antenna, ask, "Compared to what?" Isotropic refers to a theoretical point source that radiates in all directions. The term "dBi" refers to gain in decibels[47] compared to an isotropic point source.

Verticals A vertical antenna is omnidirectional, radiating and receiving equally in all directions around the base. It has a low radiation angle, meaning the signal stays close to the horizon. You would not use a vertical to make contact through an overhead satellite.

The signal doesn't shoot straight up or straight down. Therefore, although omnidirectional around the horizon, the vertical does exhibit 3 dBi gain, and 3 dB is twice the power of a theoretical point source. Stacked verticals can exhibit even higher gain by concentrating the signal closer to the horizon.

Dipoles The dipole is bi-directional and enhances signals off the two sides in a figure-eight pattern centered on the antenna's center feed-point. Hardly anything radiates off the ends. This squeezes the radiated power into a narrower area, increasing the signal in and from the favored directions. The dipole has 6 dBi gain to the sides, four times the point source, and double the 3 dBi vertical.

If the dipole is horizontal, ground reflection fills in the pattern, and the antenna becomes more omnidirectional. The ground's effect is more

---

[47] Decibels are logarithmic. 10 dB is ten times the power. 3dB is double.

ANTENNAS

pronounced with a lower antenna. Below a half wavelength, the antenna is very close to omnidirectional. That would be about 66 feet on 40 meters (7 MHz).

There is a minimal ground effect on UHF/VHF because the wavelengths are quite short. An antenna mounted at six feet is a full wavelength high on VHF, including the 2-meter Ham band. On UHF (GMRS), the same antenna is 12 wavelengths high.

Directional Antennas Beams, Yagis,[48] or quads are the names for directional antenna designs focusing energy in one direction. The terms "beam" and "Yagi" are used interchangeably.

Beams can significantly increase the range. They typically exhibit 6 dBd gain or 6 decibels above a dipole. That would be 4 times the transmitter's power going in the pointed direction. Since the dipole is 3 dB above an isotropic source, the beam is 6 + 3 or 9 dBi above the same isotropic source (8 times). When comparing antenna advertising claims, pay attention to the scale. An antenna claiming 6 dBi is the same as one claiming 3 dBd. Retailers often cite the dBi figure because "more" looks better.

For example, a five-element Yagi (beam) claims 9 dBi gain (6 dB above a vertical at 3 dBi). That is 4 times the power of a vertical. Five watts to a beam is equivalent to a vertical at twenty watts. The Yagi is cheaper than a higher-powered radio, has the same multiplier effect on reception, and attenuates signals from other directions, reducing interference.

Yagi design starts with the humble dipole and adds one or more elements in parallel. The reflector is about 5% longer than the driven element (dipole). A

---

[48] So-named for one of the inventors, Hidetsugu Yagi.

ANTENNAS

three-element beam adds a director to the antenna's front about 5% shorter than the driven element. The size and spacing of the elements determine the antenna's gain, pattern, and bandwidth. There are online calculators to roll your own or choose a commercially made VHF/UHF version for around $50.

Directional VHF/UHF antennas are small enough to be handheld or turned with a small TV rotor. A VHF Yagi only has a wingspan of about three feet. A UHF Yagi is less than a foot wide. Mount the Yagi on a push-up painter's pole and point it using the "Armstrong" method for temporary portable operation.

RECOMMENDATIONS

- When the path is difficult, establish a fixed link on CB or GMRS between Point A and Point B by setting up a directional antenna at one or both ends
- Get the antennas as high as possible so they "see" each other; the combination of height and directivity will improve the connection
- If there are obstructions between two stations, try pointing the antennas to bounce the signal off a building or another obstacle to go around the obstruction
- Use an NVIS antenna on HF frequencies (requires a Ham license).

FEEDLINE

A rubber-ducky antenna connects directly to the transceiver. FRS antennas are not removable, but other services allow substitute antennas. An antenna that is not affixed to the radio needs a feed line. The most common feed line is coaxial cable or "coax."

Coax consists of a center conductor, insulation, a shield, and an outer insulation layer. Each dimension determines the coax's impedance, loss, and power-handling capability.

Match the antenna's impedance to the feed line and the radio. Matching impedances transfer the maximum power. For most of our applications, that impedance is 50 ohms, and so is most coax. Almost all coax can handle 100 watts, so your primary concern is the loss inherent in the cable.

Loss Loss is measured in decibels (dB) per 100 feet. Decibels express a logarithmic ratio of power. Plus 3 decibels is a doubling of power, plus another 3dB (6 dB) is doubling again or four times the power. Likewise, -3 decibels is half the power, and -6dB is one-quarter the power. If you put 5 watts into a line with a 6dB loss, only a little more than a watt makes it to the antenna.

Coax loss increases rapidly with frequency and higher SWR, so having the proper coax is essential on VHF/UHF. Coax suitable for HF, like RG58,[49] has 12dB loss per hundred feet at 400 MHz (UHF). Only one-sixteenth of your power gets to the antenna. By contrast, LMR400 has a 2.7dB loss, and about one-half of the power makes the trip.

---

[49] RG58 is a common designation for a popular coax type used on HF. Manufacturers may have their own naming system so consult the charts.

FEEDLINE

Moisture Water is the number one enemy of coax. To run a coax cable underground, use cable rated for that service, or water will penetrate the outer insulation. Moisture intrusion causes additional losses, and the effect is so gradual you won't notice.

Moisture can come in through the connectors. Seal the end of the coax and any connectors to keep out water. Use good quality electrical tape and cover it with Coax-Seal, a putty-like tape. Then cover the Coax-Seal with another layer of electrical tape. The first layer of tape makes it easier to remove the sticky Coax-Seal later. Properly installed coax should last many years.

Finally, moisture can intrude because of animal damage. Squirrels chew coax and penetrate the outer insulation. I check my lines whenever I can and usually find evidence of squirrel damage. I can't imagine what they find so tasty.

RECOMMENDATIONS

- Select the coax type with an acceptable loss number at your intended frequency and budget
- Keep the coax lines short; if the loss is 12dB for 100 feet, it will be half or 6dB for 50 feet
- If you have a few feet left over, don't obsess over cable length, as a few extra feet won't be noticeable.

MOBILE OPERATION

CB, GMRS, and Ham radios lend themselves to mobile operation. A car-mounted radio can also serve as a fixed base station, taking advantage of the car's battery and engine power. The car will idle for about 72 hours on a full gas tank. Operating an hour a day, it would last over two months.

Some transceivers designed for home or mobile operations have a small detachable control head that fits into the cabin. Mount the larger transceiver in the trunk or under a seat. Secure the transceiver well so it does not become a flying missile during a car accident or slide around and create a distraction. Laying a radio on the passenger seat is a dangerously bad idea.

Here is an HF/VHF/UHF transceiver in the trunk.

I was fortunate there was a bracket in the spare tire well, and I used it to secure the radio's mobile mount.

Power Run the power cable directly to the battery to reduce electrical noise and provide maximum power.

MOBILE OPERATION

Remember the earlier advice to use a short and stout wire (#10 or larger) to avoid voltage drop. A 100-watt radio draws close to 20 amps on transmit, so don't use a cigarette lighter outlet for power.

In my case, the battery was also in the trunk, making the connection easy. A front-mounted battery will require the power wire to run through the firewall. Look for a gasket or rubber boot. Use #8 wire for longer runs and protect the wire from chafing anywhere it touches metal.

Install a fuse on both the power cable's positive and negative sides for maximum protection against overload and fire. Most factory-supplied power cables will include fusing.

Wires Microphone, speaker, and control wires went under the trunk floor mat, and I temporarily removed the back seat to pull them from the trunk to the cabin. Route the cables to the front by tucking them under the door trim for a neat and clean install. An Internet search yielded several videos with complete instructions on removing the back seat.

The control head is secured with Scotch brand "Permanent Double-Sided Clear Mounting Tape." It is very sticky but removes without leaving residue. This works best if you use a small lip on the dashboard to support the weight. There are also mountings designed to fit in a cup-holder.

MOBILE OPERATION

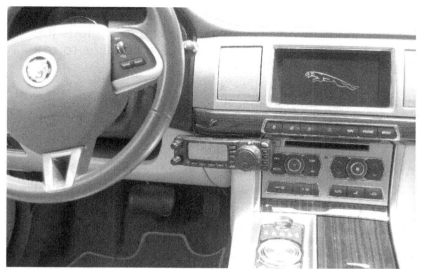

The control head is on the dashboard to the right of the steering wheel.

The speaker looks factory installed.

The next component was the speaker, and I was at a loss until I looked in the rear-view mirror and noticed the child-seat restraint on the rear deck in the middle of the back seat. Two zip-ties hold the speaker

MOBILE OPERATION

bracket securely to the restraint clip. The wire goes behind the seat and to the trunk, where it plugs into the radio. I used zip ties to attach a speaker to the rear headrest in a different car.

Before beginning an installation, sit in your car and look around. I would never have thought of the speaker mount until it jumped out at me.

Antenna Mobile antennas are verticals. They may not be a full quarter-wave long, but an inserted coil electrically lengthens the antenna. The car acts as the "other" half of the antenna. This requires a good electrical connection to the car body. I use an antenna mount that attaches to the trunk lid. Many trunk hinges ride on fiberglass bearings that insulate them from the rest of the car. A strap across the hinge connecting the lid and car body corrects that problem.

I call this "the cat's meow."
The antenna is on the left side of the trunk.

LEARNING MORSE CODE

If I said I would teach you a foreign language and all you needed to know was 26 words and count to 10, would you think it was impossible? Probably not.

Boy Scouts was my first experience with Morse code. Unfortunately, I learned it wrong and had to relearn it as a Ham. What went wrong?

In Boy Scouts, we memorized Morse as dots and dashes. The first exposure was visual, not audible, and there was a chart to look up the letters. What's wrong with that? Your brain has to convert what it sees or hears into dots and dashes and then translate those into letters. It becomes a multi-step process. It is like translating from English to Spanish to get to French.

I thought I understood it until someone sent Morse code by flashlight. It was blinking light to dots and dashes to letters. It was even worse when sent by waving a flag. I was lost.

Radio Morse is audible. Learn to recognize the sound and not memorize dots and dashes. "A" is not dot-dash. It is not even dit-dah. It is the sound of dit-dah. Learning the sound eliminates all the in-between translations.

For the same reason, do not learn that "A" sounds like "Ah-pull" or that the letter "A" has a short line and a long line. I've seen pictograms patterned after children's alphabet blocks. Ignore all these "helpers." Gimmicks introduce additional mental steps requiring we go from English to Spanish to French to get to Russian.

Koch method Learn Morse by hearing one or two letters at a time until the sound and the letter connect

LEARNING MORSE CODE

immediately. Then go on to the next letters and build. This procedure is called the Koch method.

Farnsworth Timing Another impediment to my learning was the way we sent Morse. At slow speeds, "A" became the sound made by diiiiiiit-daaaaaaaaaah. Then, the following letter came immediately with no time in between for the brain to work.

Farnsworth timing is the modern way and sends the letters at least 20 words per minute,[50] so each letter has one distinct sound. Dits and dahs may form a letter but don't listen to hear individual elements. The letter is one sound.

To slow down the pace, Farnsworth increases the space between letters and words. Increasing space does two things. It reinforces a single sound as the letter and gives the brain extra time between letters and words to work the translation. Send using Farnsworth timing because that is how the other guy learned. I set my keyer around 22-26 words per minute and slow down by increasing spaces.

After learning the letters, you will recognize words. When you read, you do not see letters; your brain jumps to the word. "Dog" is not "D-O-G." The same happens with Morse Code. Learn to recognize your call sign, RST, 5NN,[51] TU,[52] 73,[53] and other common "words" without thinking about the individual letters or the elements that make up the letters.

It takes practice. In the beginning, listen to a code practice CD or audio file. K7QO offers a free course download on his website, K7QO.net. G4FON has a

---

[50] Words per minute (WPM) is based on a 5 letter word. "Paris" is an often-used standard. Send "Paris" 20 times in a minute for 20 WPM.
[51] Short for 599, a signal report.
[52] Thank you.
[53] Best regards.

Koch trainer at G4FON.net. There are other sources, as well.

Once you get the letters down, listen to QSOs[54] for the standard QSO pattern. Then, GET ON THE AIR! There is no better practice than making actual contacts. Real QSOs are exciting and won't seem like practice.

On-the-air practice is best.

FISTS CW Club and Straight Key Century Club (SKCC) promote CW, and you will find slow CW on the frequencies suggested on their web pages. Both FISTS CW Club and SKCC assign a member number to exchange with other members and collect awards. It is a fun challenge. Suggested frequencies to find slower CW include:

3.550 – 3.600 MHz   21.055 – 21.200 MHz
7.055 – 7.125 MHz   28.055 – 28.060 MHz
14.055 – 14.060 MHz

When starting on CW, concentrate on the QSO trinity of the signal report, location, and name. The pattern is always the same, so you know what to expect. Progress to more complex conversations with time.

---

[54] QSO is a contact or conversation.

LEARNING MORSE CODE

Try to match the speed of the other guy but if you can't, "QRS" means "slow down," and "QRQ" means "speed up."

For more serious practice, tune in to the W1AW Code Practice Sessions. These texts from *QST* magazine are harder to copy because the words are longer and not as predictable.[55] ARRL offers code proficiency certificates that will look impressive on your wall.

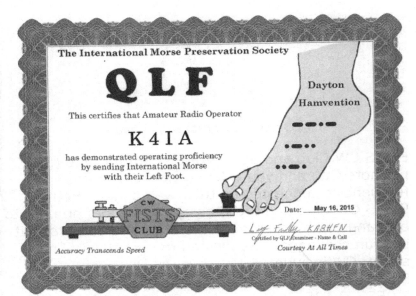

When someone has a lousy fist[56], the joke is, "QLF? Are you sending with your left foot?" I earned this certificate for demonstrating proficiency while sending using my left foot.

---

[55] http://www.arrl.org/w1aw-operating-schedule
[56] "Fist" refers to the operator's style of sending; his "accent" when sending CW.

TROUBLESHOOTING

In the scientific method, parsimony is an epistemological, metaphysical, or heuristic preference, not an irrefutable principle of logic or a scientific result. Occam's razor (law of parsimony) would suggest that scientists accept the simplest possible theoretical explanation for existing data.

Radio troubleshooting follows the same principle. The problem is often something simple. That is good because modern radios are almost impossible to work on without training and sophisticated test equipment. Radio parts are minuscule and mounted on layers of stacked circuit boards. Stay out of the radio unless you know what you are doing.

Start troubleshooting by asking, "What changed?" Did you recently change the power supply, antenna, or microphone? Did any settings change? Go back and undo the change. Did you move the equipment? Perhaps a cable disconnected or got plugged back into the wrong place. Was there a storm that blew down the antenna or damaged the feed line?

What can go wrong? Most problems are operator errors.

EQUIPMENT FAILURE

If the radio doesn't turn on, check the power and associated cables:
- Is the power source "on?"
- Check the power connectors to make sure one isn't loose or reversed
- Use a multi-meter to measure the voltage on the power source; it should read around 14 volts
- If the power source is good, check for blown fuses in the power cord; a blown fuse may show

TROUBLESHOOTING

- blackened glass, or you won't be able to see the thin wire that connects the two ends
- Use a multi-meter to check the resistance from one end of the fuse to the other (with the fuse removed from the holder)
- If the resistance is infinite, the fuse is a blown (open); try to deduce why it blew
- If something caused the radio to draw excessive current, were the power wires crossed? Is the antenna mismatched? Put in another fuse of the same size, and hope it was just a fluke; don't upsize the fuse, or you risk equipment damage or fire
- If the fuse is good, check the voltage at the radio end of the cable
- If there is no voltage at the radio, disconnect the power and check the power cable's resistance; if it is infinite, the cable or connector is broken and needs to be patched or replaced
- Most problems are with the connector and, specifically, where the wire attaches to the connector
- If the radio turns off when transmitting, suspect a voltage drop caused by a too small or too long power cable or a weak battery.

I CAN'T HEAR ANYBODY, OR THEY CAN'T HEAR ME

If the radio turns on, but you don't hear any signals:
- Is the radio locked in transmit mode? Some microphone PTT switches have a lock
- Is the RF or audio gain turned down?
- Is the squelch control turned up too high for signals to trigger?
- Is the antenna connection loose? Unscrew it and re-attach to be sure
- You could have a faulty antenna connector
- Measure the antenna feed line with a multi-meter; set it to ohms; there should be no

TROUBLESHOOTING

resistance along the cable and infinite resistance between the core and the shield
- Some antennas have a coil across the core and ground; they will exhibit a short when connected to an ohm-meter; best to disconnect the antenna from the feed line and measure the feed line alone
- Try a different antenna and cable
- Are the frequency, offset, and tones set correctly when using a repeater? One or all is usually the problem
- Are the tones set correctly when using an FRS or GMRS radio?

If you can hear signals, but no one hears you transmitting:
- Is the radio in the wrong mode? The microphone doesn't work if the radio is in a non-voice mode like CW or Digital
- Is the microphone gain turned down?
- Is the microphone cable frayed, loose, or not attached properly?
- Try a different microphone
- Attach an SWR or power meter and see if it shows any signal going out
- Ask a friend to listen
- Are the offset and tones set correctly?
- Check that the radio is not in low power, test, or standby mode
- Is there an appropriately sized power supply and wire? If the wire is too long or too thin, a voltage drop could affect the transmitter; the receiver sounds fine because it draws little current and has less voltage drop.

SUMMARY AND RECOMMENDATIONS

This book's purpose was to show you The Easy Way to prepare for emergency and prepper communications. Here is a summary of The Easy Way methods:

- Prepare now
- Get involved with a local radio club
- Find an Elmer or two
- Listen, listen, listen, and learn
- Get on the air no matter how modest your station or antenna
- Don't despise humble beginnings
- Learn and practice the 3-3-3 rule.

My recommendations for a station would be:

- AM/FM/Shortwave/NOAA portable radio
- CB and GMRS radio for local communication
- At least three power sources for all equipment
- Inverter type generator
- Battery backup (20 amp-hour or more) with at least 100 watts of solar cells
- Amateur Radio Technician License
- Portable HT and more powerful base station, either Amateur or GMRS
- Directional antenna and a way to elevate your antenna at least 15 feet
- For the more advanced prepper, Amateur Radio General Class license and HF radio station.

73/DX
Buck
K4ia

APPENDIX A ANTENNA LENGTHS

Values are for each half of a dipole based on 234/frequency = length in feet. Cut longer and adjust. To adjust, note the actual and target frequency of the lowest SWR. Add or subtract the difference on each half of your antenna. To move lower, add wire. To move higher, wrap it back. Bold frequencies are in the approximate Amateur bands.

MHz	Feet	Inches	MHz	Feet	Inches
	80 M			**30 M**	
3.2	73.2	878	10	23.4	281
3.3	71	851	10.05	23.3	279
3.4	68 5/6	826	**10.1**	23.2	278
3.5	66 6/7	802	**10.15**	23	277
3.6	65	780	10.2	23	275
3.7	63.25	759	10.25	22.8	274
3.8	61.6	739	10.3	22.7	273
3.9	60	720	10.35	22.6	271
4.0	58.5	702	10.4	22.5	270
4.1	57	685	10.45	22.4	269
4.2	55 .7	669	10.5	22.3	267
	40 M			**20 M**	
6.7	35	419	13.9	15.8	202
6.8	34.4	413	**14.0**	16.7	201
6.9	34	407	**14.1**	16.6	199
7.0	33 3/7	401	**14.2**	16.5	198
7.1	33	395	**14.3**	16.3	196
7.2	32.5	390	**14.4**	16.25	195
7.3	32	385	14.5	16.1	194
7.4	31 5/8	379	14.6	16	192
7.5	31.2	374	14.7	16	191
7.6	30.8	369	14.8	15.8	190
7.7	30.4	365	14.9	15.7	188

TROUBLESHOOTING

MHz	Feet	Inches	MHz	Feet	Inches
	17 M			**10 M**	
17	13.75	165	27.5	8.5	102
17.5	13.4	160	28	8.3	100
18	13	156	**28.2**	8.3	100
18.068	13	155	**28.4**	8.25	99
18.168	12.8	155	**28.6**	8.2	98
18.25	12.8	154	**28.8**	8.1	98
18.35	12.75	153	**29**	8	97
18.45	12.6	152	**29.2**	8	96
18.55	12.6	151	**29.4**	8	96
18.65	12.6	151	**29.6**	8	96
18.75	12.5	150	30	7.8	94
	15 M				
20.8	11.25	135		**VHF**	
20.9	11.2	134	**144**	1.625	20
21.0	11.1	134	**148**	1.6	19
21.1	11	133			
21.2	11	132		**UHF**	
21.3	11	132	**440**	.5	6
21.45	11	131	**448**	.5	6
21.55	10.8	130			
21.65	10.8	130		**GMRS**	
21.75	10.75	129	462	.5	6
21.85	10.7	129	467	.5	6
	12 M				
24	9.75	117			
24.849	9.4	113			
24.94	9.4	113			
25.1	9.3	112			
25.2	9.3	111			
25.3	9.25	111			

APPENDIX B 50,000 WATT AM STATIONS

The most power allowed on the AM broadcast band in the United States is 50,000 watts. You can often hear these mega-stations at night, coast-to-coast. There is no guarantee they will operate at full power in an emergency, but you may get information from a part of the country not affected by your local conditions. See how many you can hear.

The list is in ascending order by frequency.

Station	Frequency	Location
WFLF	540 kHz	Orlando, Florida
KMJ	580 kHz	Fresno, California
WTCM	580 kHz	Traverse City, Michigan
WTMJ	620 kHz	Milwaukee, Wisconsin
WGST	640 kHz	Atlanta, Georgia
KFI	640 kHz	Los Angeles, California
WWJZ	640 kHz	Mount Holly, New Jersey
WNNZ	640 kHz	Westfield, Massachusetts
KENI	650 kHz	Anchorage, Alaska
WSM	650 kHz	Nashville, Tennessee
WLFJ	660 kHz	Greenville, South Carolina
WFAN	660 kHz	New York
KTNN	660 kHz	Window Rock, Arizona, Navajo Nation
KBOI	670 kHz	Boise, Idaho
WSCR	670 kHz	Chicago, Illinois
KLTT	670 kHz	Commerce City, Colorado
WWFE	670 kHz	Miami, Florida
WCBM	680 kHz	Baltimore, Maryland
WRKO	680 kHz	Boston, Massachusetts
WCNN	680 kHz	North Atlanta, Georgia
WPTF	680 kHz	Raleigh, North Carolina
KKYX	680 kHz	San Antonio, Texas

50,000 WATT AM STATIONS

KNBR	680 kHz	San Francisco, California
WJOX	690 kHz	Birmingham, Alabama
WOKV	690 kHz	Jacksonville, Florida
WLW	700 kHz	Cincinnati, Ohio
KALL	700 kHz	North Salt Lake City, Utah
KXMR	710 kHz	Bismark, North Dakota
KSPN	710 kHz	Los Angeles, California
WAQI	710 kHz	Miami, Florida
WOR	710 kHz	New York
KIRO	710 kHz	Seattle, Washington
KEEL	710 kHz	Shreveport, Louisiana
WGN	720 kHz	Chicago, Illinois
KDWN	720 kHz	Las Vegas, Nevada
WGCR	720 kHz	Pisgah Forest, North Carolina
KBRT	740 kHz	Costa Mesa, California
KNFL	740 kHz	Fargo, North Dakota
KTRH	740 kHz	Houston, Texas
WYGM	740 kHz	Orlando, Florida
KCBS	740 kHz	San Francisco, California
KRMG	740 kHz	Tulsa, Oklahoma
KFQD	750 kHz	Anchorage, Alaska
WSB	750 kHz	Atlanta, Georgia
KERR	750 kHz	Polson, Montana
KXTG	750 kHz	Portland, Oregon
WJR	760 kHz	Detroit, Michigan
KTKR	760 kHz	San Antonio, Texas
KFMB	760 kHz	San Diego, California
KDSP	760 kHz	Thornton, Colorado
KKOB	770 kHz	Albuquerque, New Mexico
KCBC	770 kHz	Manteca, California
WABC	770 kHz	New York
KTTH	770 kHz	Seattle, Washington
WBBM	780 kHz	Chicago, Illinois
KKOH	780 kHz	Reno, Nevada
WHB	810 kHz	Kansas City, Missouri

50,000 WATT AM STATIONS

WSJC	810 kHz	Magee, Mississippi
KGO	810 kHz	San Francisco, California
WKVM	810 kHz	San Juan, Puerto Rico
WGY	810 kHz	Schenectady, New York
KGNW	820 kHz	Burien, Washington
WBAP	820 kHz	Ft Worth, Texas
KUTR	820 kHz	Taylorsville, Utah
WTRU	830 kHz	Kernersville, North Carolina
WCCO	830 kHz	Minneapolis, Minnesota
KLAA	830 kHz	Orange, California
KFLT	830 kHz	Tucson, Arizona
WCRN	830 kHz	Worcester, Massachusetts
WCEO	840 kHz	Columbia, South Carolina
WHAS	840 kHz	Louisville, Kentucky
KXNT	840 kHz	North Las Vegas, Nevada
WXJC	850 kHz	Birmingham, Alabama
WEEI	850 kHz	Boston, Massachusetts
WKNR	850 kHz	Cleveland, Ohio
KOA	850 kHz	Denver, Colorado
KICY	850 kHz	Nome, Alaska
WTAR	850 kHz	Norfolk, Virginia
KTRB	860 kHz	San Francisco, California
KPAM	860 kHz	Troutdale, Oregon
KRLA	870 kHz	Glendale, California
WWL	870 kHz	New Orleans, Louisiana
KRVN	880 kHz	Lexington, Nebraska
WCBS	880 kHz	New York
KLRG	880 kHz	Sheridan, Arkansas
WBAJ	890 kHz	Blythewood, South Carolina
WLS	890 kHz	Chicago, Illinois
KTIS	900 kHz	Minneapolis, Minnesota
WFDF	910 kHz	Farmington Hills, Michigan
KFIG	940 kHz	Fresno, California
WMAC	940 kHz	Macon, Georgia
WINZ	940 kHz	Miami, Florida

50,000 WATT AM STATIONS

Station	Frequency	Location
WWJ	950 kHz	Detroit, Michigan
KJR	950 kHz	Seattle, Washington
WTEM	980 kHz	Washington, D.C.
WDYZ	990 kHz	Orlando, Florida
WNTP	990 kHz	Philadelphia, Pennsylvania
WMVP	1000 kHz	Chicago, Illinois
KOMO	1000 kHz	Seattle, Washington
WTZA	1010 kHz	Atlanta, Georgia
WJXL	1010 kHz	Jacksonville Beach, Florida
WINS	1010 kHz	New York
WHFS	1010 kHz	Seffner, Florida
KXEN	1010 kHz	St. Louis, Missouri
KIHU	1010 kHz	Tooele, Utah
KTNQ	1020 kHz	Los Angeles, California
KDKA	1020 kHz	Pittsburgh, Pennsylvania
KMMQ	1020 kHz	Plattsmouth, Nebraska
KCKN	1020 kHz	Roswell, New Mexico
WBZ	1030 kHz	Boston, Massachusetts
KTWO	1030 kHz	Casper, Wyoming
KCTA	1030 kHz	Corpus Christi, Texas
WWGB	1030 kHz	Indian Head, Maryland
WCTS	1030 kHz	Maplewood, Minnesota
WGSF	1030 kHz	Memphis, Tennessee
KDUN	1030 kHz	Reedsport, Oregon
WDRU	1030 kHz	Wake Forest, North Carolina
WPBS	1040 kHz	Atlanta, Georgia
WHO	1040 kHz	Des Moines, Iowa
WEPN	1050 kHz	New York
KTCT	1050 kHz	San Mateo, California
KRCN	1060 kHz	Longmont, Colorado
WQOM	1060 kHz	Natick, Massachusetts
WLNO	1060 kHz	New Orleans, Louisiana
KYW	1060 kHz	Philadelphia, Pennsylvania
WKNG	1060 kHz	Talapoosa, Georgia
WIXC	1060 kHz	Titusville, Florida

50,000 WATT AM STATIONS

WAPI	1070 kHz	Birmingham, Alabama
WNCT	1070 kHz	Greenville, North Carolina
WFNI	1070 kHz	Indianapolis, Indiana
WFLI	1070 kHz	Lookout Mountain, Tennessee
KNX	1070 kHz	Los Angeles, California
WDIA	1070 kHz	Memphis, Tennessee
WCSZ	1070 kHz	Sans Souci, South Carolina
WKAT	1080 kHz	Coral Gables, Florida
KRLD	1080 kHz	Dallas, Texas
WTIC	1080 kHz	Hartford, Connecticut
WWNL	1080 kHz	Pittsburgh, Pennsylvania
KFXX	1080 kHz	Portland, Oregon
KMXA	1090 kHz	Aurora, Colorado
WBAL	1090 kHz	Baltimore, Maryland
KAAY	1090 kHz	Little Rock, Arkansas
KFNQ	1090 kHz	Seattle, Washington
WTAM	1100 kHz	Cleveland, Ohio
WZFG	1100 kHz	Dilworth, Minnesota
KNZZ	1100 kHz	Grand Junction, Colorado
KFNX	1100 kHz	Phoenix, Arizona
KFAX	1100 kHz	San Francisco, California
WBT	1110 kHz	Charlotte, North Carolina
KVTT	1110 kHz	Mineral Wells, Texas
KFAB	1110 kHz	Omaha, Nebraska
KRDC	1110 kHz	Pasadena, California
KPNW	1120 kHz	Eugene, Oregon
KMOX	1120 kHz	Saint Louis, Missouri
WDFN	1130 kHz	Detroit, Michigan
WISN	1130 kHz	Milwaukee, Wisconsin
KTLK	1130 kHz	Minneapolis, Minnesota
WBBR	1130 kHz	New York
KWKH	1130 kHz	Shreveport, Louisiana
WQBA	1140 kHz	Miami, Florida
WRVA	1140 kHz	Richmond, Virginia
KHTK	1140 kHz	Sacramento, California

50,000 WATT AM STATIONS

Station	Frequency	Location
KEIB	1150 kHz	Los Angeles, California
WYLL	1160 kHz	Chicago, Illinois
WCRT	1160 kHz	Donelson, Tennessee
KSL	1160 kHz	Salt Lake City, Utah
KJNP	1170 kHz	North Pole, Alaska
KCBQ	1170 kHz	San Diego, California
KLOK	1170 kHz	San Jose, California
KFAQ	1170 kHz	Tulsa, Oklahoma
WWVA	1170 kHz	Wheeling, West Virginia
KGOL	1180 kHz	Humble, Texas
KOFI	1180 kHz	Kalispell, Montana
WJNT	1180 kHz	Pearl, Mississippi
WHAM	1180 kHz	Rochester, New York
KYES	1180 kHz	Rockville, Minnesota
KERN	1180 kHz	Wasco-Green Acres, California
KFXR	1190 kHz	Dallas, Texas
WOWO	1190 kHz	Fort Wayne, Indiana
WCRW	1190 kHz	Leesburg, Virginia
KEX	1190 kHz	Portland, Oregon
WXKS	1200 kHz	Newton, Massachusetts
WAXA	1200 kHz	Pine Island Center, Florida
WOAI	1200 kHz	San Antonio, Texas
WMUZ	1200 kHz	Taylor, Michigan
KFNW	1200 kHz	West Fargo, North Dakota
WJNL	1210 kHz	Kingsley, Michigan
WPHT	1210 kHz	Philadelphia, Pennsylvania
WHKW	1220 kHz	Cleveland, Ohio
WXYT	1270 kHz	Detroit, Michigan
KFLC	1270 kHz	Fort Worth, Texas
WADO	1280 kHz	New York, New York
WNQM	1300 kHz	Nashville, Tennessee
WJNJ	1320 kHz	Jacksonville, Florida
KPXQ	1360 kHz	Glendale, Arizona
KMNY	1360 kHz	Hurst, Texas
WQLL	1370 kHz	Pikesville, Maryland

50,000 WATT AM STATIONS

KRKO	1380 kHz	Everett, Washington
KZQZ	1430 kHz	St. Louis, Missouri
KTNO	1440 kHz	University Park, Texas
WWNN	1470 kHz	Pompano Beach, Florida
WLQV	1500 kHz	Detroit, Michigan
KSTP	1500 kHz	Saint Paul, Minnesota
WFED	1500 kHz	Washington, D.C.
WMEX	1510 kHz	Boston, Massachusetts
WWBC	1510 kHz	Cocoa, Florida
WLAC	1510 kHz	Nashville, Tennessee
KGA	1510 kHz	Spokane, Washington
WWKB	1520 kHz	Buffalo, New York
KOKC	1520 kHz	Oklahoma City, Oklahoma
KQRR	1520 kHz	Oregon City, Oregon
KKXA	1520 kHz	Snohomish, Washington
WCKY	1530 kHz	Cincinnati, Ohio
KGBT	1530 kHz	Harlingen, Texas
WYMM	1530 kHz	Jacksonville, Florida
KFBK	1530 kHz	Sacramento, California
WDCD	1540 kHz	Albany, New York
KMPC	1540 kHz	Los Angeles, California
WNWR	1540 kHz	Philadelphia, Pennsylvania
KXEL	1540 kHz	Waterloo, Iowa
KRPI	1550 kHz	Ferndale, Washington
WLOR	1550 kHz	Huntsville, Alabama
WAZX	1550 kHz	Smyrna, Georgia
KUAZ	1550 kHz	Tucson, Arizona
KKOV	1550 kHz	Vancouver, Washington
WFME	1560 kHz	New York
WJFK	1580 kHz	Morningside, Maryland
KQFN	1580 kHz	Phoenix, Arizona
KBLA	1580 kHz	Santa Monica, California
WMQM	1600 kHz	Lakeland, Tennessee

APPENDIX C GMRS/FRS FREQUENCIES

Frequencies are in megahertz (MHz).

Channel	Type	Frequency	FRS Power	GMRS Power
1	FRS/GMRS	462.5625	2W	5W
2	FRS/GMRS	462.5875	2W	5W
3	FRS/GMRS	462.6125	2W	5W
4	FRS/GMRS	462.6375	2W	5W
5	FRS/GMRS	462.6625	2W	5W
6	FRS/GMRS	462.6875	2W	5W
7	FRS/GMRS	462.7125	2W	5W
8	FRS/GMRS	467.5625	0.5W	0.5W
9	FRS/GMRS	467.5875	0.5W	0.5W
10	FRS/GMRS	467.6125	0.5W	0.5W
11	FRS/GMRS	467.6375	0.5W	0.5W
12	FRS/GMRS	467.6625	0.5W	0.5W
13	FRS/GMRS	467.6875	0.5W	0.5W
14	FRS/GMRS	467.7125	0.5W	0.5W
15	FRS/GMRS	462.5500	2W	50W
16	FRS/GMRS	462.5750	2W	50W
17	FRS/GMRS	462.6000	2W	50W
18	FRS/GMRS	462.6250	2W	50W
19	FRS/GMRS	462.6500	2W	50W
20	FRS/GMRS	462.6750	2W	50W
21	FRS/GMRS	462.7000	2W	50W
22	FRS/GMRS	462.7250	2W	50W
RPT15	GMRS	467.5500	-	50W
RPT16	GMRS	467.5750	-	50W
RPT17	GMRS	467.6000	-	50W
RPT18	GMRS	467.6250	-	50W
RPT19	GMRS	467.6500	-	50W
RPT20	GMRS	467.6750	-	50W
PRT21	GMRS	467.7000	-	50W
RPT22	GMRS	467.7250	-	50W

Channels RPT15 – 22 are for repeaters

APPENDIX D PREPPER RADIO FREQUENCIES

Description	Frequency MHz	Mode	Tone May vary
Ham 80 Mtr	3.818	LSB	
Ham 60 Mtr	5.357	USB	
Marine Emergency	6.215	USB	
Ham 40 Mtr	7.242	LSB	
Ham 20 Mtr	14.242	USB	
Ham Maritime	14.300	USB	
CB CH 3	26.985	AM	
CB CH 4	27.005	AM	
CB CH 19	27.185	AM	
CB CH 36	27.365	USB	
CB CH 27	27.375	USB	
Ham 10 Mtr	28.305	USB	
Aircraft Emergency	121.500	AM	
Ham 2 Mtr	146.420	FM	100
Ham 2 Mtr Calling	146.520	FM	
Ham 2 Mtr	146.550	FM	151.4
MURS Primary	151.940	NarrowFM	67
MURS Prepper	154.570	FM	67
Search and Rescue	151.160	FM	127.3
Marine Safety	156.800	FM	
Marine Prepper	156.625	FM	
Ham UHF Calling	446.000	FM	100
Ham UHF Prepper	446.030	FM	100
FRS 1 Calling	462.562.5	NFM	67
FRS 2 Chat	462.587.5	NFM	67
FRS 3 Prepper	462.612.5	NFM	67
FRS 4 Prepper	462.637.5	NFM	67
GMRS Ch 17 Prepper	462.600		118.8
GMRS Ch 20 Travel	462.675		131.8

APPENDIX E BATTERY CAPACITY

The following chart assumes LiPO4 Battery 20% receive and 80% transmit duty cycle.
Line up the transmit power and desired runtime, then read the recommended battery capacity in watt hours. Divide by 12 to get amp hours.

Trans Watts	Recv Watts	Avg* Watts	Watt Hours**	Run Hours ***
5	5	5	36	7.2
		5	54	10.8
		5	72	14.4
		5	108	21.6
		5	144	28.8
10	5	6	36	6
		6	54	9
		6	72	12
		6	108	18
		6	144	24
		6	180	30
20	5	8	72	9
		8	108	13.5
		8	144	18
		8	180	22.5
		8	240	30
25	5	9	72	8
		9	108	12
		9	144	16
		9	180	20
		9	240	26.7

BATTERY CAPACITY

50	5	14	108	7.7
		14	144	10.3
		14	180	12.8
		14	240	17.1
75	5	19	108	5.6
		19	144	7.5
		19	180	9.4
		19	240	12.6
100	5	24	144	6
		24	180	7.5
		24	240	10
		24	360	15
200	5	44	240	5.4
		44	480	10.9
		44	720	16.3
		44	1200	27.2

* Average watts is the calculated average based on 20% transmit time, and 80% receive.

** 36 watt hours divided by 12 volts = 3 amp hours.

*** Increase this figure by 60% or decrease runtime by 60% when using lead-acid batteries. They do not deliver full power for the full rating time.

APPENDIX F POWER CABLE

EXTENSION CORDS AT 120 VOLTS

25 - 50 feet:
- 16 Gauge[57] for 1-13 Amps
- 14 Gauge for 14-15 Amps
- 12-10 Gauge for 16-20 Amps

100 feet:
- 16 Gauge for 1-10 Amps
- 14 Gauge for 11-13 Amps
- 12 Gauge for 14-15 Amps
- 10 Gauge for 16-20 Amps

RECOMMENDED LENGTH AND AMPERAGE FOR WIRE WITH LESS THAN 2% VOLTAGE DROP AT 12 VOLTS

GAUGE	AMPS	5	10	15	20	25	30
20		4	2	1			
28		7	3	2	1		
16		11	6	4	3	2	2
14		18	9	6	5	4	3
12		29	15	10	7	6	5
10		47	23	15	12	9	8
8		74	37	25	19	15	12

Note: Wire length is the total length, in feet, of both the positive and negative lead.

---

[57] "Gauge" is a measure of thickness. Lower gauge is thicker wire.

APPENDIX G POWER REQUIREMENTS

Appliances and electrical equipment often have labels listing their exact power consumption. This table is for estimating.

APPLIANCE	RUN WATTS	SURGE ADDS
Central AC	3,800	11,400
Coffee pot	1,000	
Computer monitor	250	
Desktop computer	100	350
Deep freezer	450	900
Dishwasher	1,500	1,500
Dryer (electric)	5,400	6,750
Dryer (gas)	700	1,800
Electric heater	2,000	
Electric oven	2,150	
Electric water heater	4,000	
Furnace blower	800	2,400
Heat pump	4,700	4,500
Hot plate	1,200	1,725
Laptop computer	50	
Microwave	650	800
Radio transvr 100w	840	
Refrigerator/freezer	700	2,200
Sump pump	1,050	2,150
Table fan	200	200
TV 49" LED	85	
Vacuum cleaner	200	200
Washing Machine	1,150	2,250
Well pump	1,000	2,100
Window AC	1,200	3,600

INDEX

3-3-3 rule, 99
Accessory Jack, 74
AF Gain, 67
AGC (Automatic Gain Control), 67
ALC, 74
Alternating current, 7
Amp-hour, 20, 21, 22
Amplifier Keyer, 74
Analyzer, Antenna, 58
Anderson Power Poles, 27, 59
Antenna Selector, 67
Antenna Tuner, 57, 62, 63, 68, 111
Attenuation, 67
Balun, 59
Band Switch, 68
Battery, 19
Compression, 68
CSCE, 53
CTCSS, 94
CW, 10
CW Pitch, 68
CW Speed, 68
Dipole
    Humble Wire, 113
    Multi-Band, 117, 118
DTMF, 94, 95
EMP, 95, 104, 105, 106
Filter Shift, 69
Filter Width, 68
Filters, 59, 69
Frequency, 7
Frequency Entry, 68
FRN number, 40, 52

Fuse or circuit breaker, 74
Generators
    Construction, 16
    Inverter, 17
Ground
    Electrical Safety, 106
    Lightning, 106
    RF, 106, 109
Grounding lug, 74
HT, 41, 86, 90, 91, 105, 134
Inverter, 17, 23, 27
IRLP, 95
Key or paddles, 74
Lead-acid, 18, 19
Line A, 42
Line in and Line Out, 75
Line isolator, 59
Magazines, 85
Memories, 69
Mesh networking, 97
Mic Gain, 69
Modulation, 9
Monitor, 69
Morse code, 10
Multi-Meter, 60
Noise Blanker, 69
Noise Reduction, 69
Notch, 70
NVIS, 117
Passband, 69
PBT, passband tuning, 69
Polarization, 116, 117
Power, 70

INDEX

Pre-Amp, 70
PTT, 75
QSK, 70
Receiver Incremental Tuning (RIT or Clarifier), 70
Relay, 92
Repeater, 92, 93, 94, 95
Resistance, 11, 12
Reverse, 70
RF Gain, 67, 71
RFI, 25, 109, 110, 111
Serial/USB port, 75
Simplex, 92
Speaker, 60
Spot, 71
Squelch, 71
SSB, 9, 10
SWR Meter, 61
Toroidal choke, 110
Transmit Incremental Tuning, 71
Tuning step, 71
Verticals, 115
Voltage, 11, 12
VOX, 71
  VOX Anti-Vox, 72
  VOX Delay, 71
Watts, 12
Whole-House Battery Systems, 15
Whole-House Generators, 14
Yagi, 119, 120
Zero-Beat, 68

Made in the USA
Middletown, DE
14 October 2023

40615586R00086